T0239405

Agricultural Implications of Fukushima Nuclear Accident (IV)

Tomoko M. Nakanishi • Keitaro Tanoi
Editors

Agricultural Implications of Fukushima Nuclear Accident (IV)

After 10 Years

 Springer

Editors
Tomoko M. Nakanishi
Graduate School of Agricultural
The University of Tokyo
Tokyo, Tokyo, Japan

Keitaro Tanoi
Graduate School of Agricultural and Life
The University of Tokyo
Bunkyo-ku, Tokyo, Japan

This work was supported by University of Tokyo

ISBN 978-981-19-9363-3   ISBN 978-981-19-9361-9   (eBook)
https://doi.org/10.1007/978-981-19-9361-9

This Springer imprint is published by the registered company Springer Nature Singapore Pte Ltd.
The registered company address is: 152 Beach Road, #21-01/04 Gateway East, Singapore 189721, Singapore

Foreword

The March 2011 Great East Japan Earthquake and the resulting tsunami caused an accident at TEPCO's Fukushima Daiichi Nuclear Power Station. This accident resulted in the release of radioactive materials into the surrounding areas, primarily in Fukushima Prefecture; and 10 years later, as many as 30,000 residents remain evacuated. Approximately 80% of the land area contaminated by radioactive materials was in areas related to agriculture, forestry, fisheries, and livestock. These industries were placed in a difficult situation and were forced to recover as they dealt with radioactive contamination.

Immediately after the accident at the Fukushima Daiichi Nuclear Power Plant, the Graduate School of Agricultural and Life Sciences of the University of Tokyo formed the Forum Group on Radiation Effects in Agriculture and has continued to provide support for issues relating to the accident with a focus on the affected areas. This forum group has been working to combine the diverse expertise of the Graduate School of Agricultural and Life Sciences. They are at work elucidating the large number of complex and intertwined factors faced in dealing with the radioactive contamination that affect the region. To this end, the Graduate School of Agricultural and Life Sciences has established an inter-discipline research system and has been working across disciplines to solve the relevant issues. We have established a cooperative system both with other departments within the University of Tokyo, as well as with Fukushima Prefecture and related municipalities.

The results of the efforts these faculty members have taken have been accumulated in the form of numerous scientific findings, and as such we have been holding regular debriefing sessions with the public since the incident in 2011 to discuss what we have uncovered. There have now been 16 such sessions. These findings were also published by Springer in 2013, 2016, and 2019 in the form of open access electronic books, along with the paper book versions of the findings. The electronic versions have been downloaded countless times, with the first two books already being downloaded more than 100,000 times.

Now, 10 years after the accident, we are publishing the fourth book. We hope that this helps spread awareness of the accident at the Fukushima Daiichi Nuclear Power Plant and the continuous efforts of the Graduate School of Agricultural and Life Sciences, the University of Tokyo.

Graduate School of Agricultural and   Nobuhiro Tsutsumi
Life Sciences, The University of Tokyo,
Bunkyo-ku, Tokyo, Japan

Preface

Ten years have passed since the Fukushima Nuclear Power Plant accident. In these 10 years, we summarized what we have found from the investigation of radioactivity contamination and could publish these results in three books, in 2013, 2016, and 2019, from Springer as open access books. Now, this is the fourth book of this series. We are a group of different kinds of experts in soil, plants, live stocks, mountains, radioactivity measurement, etc. We all went to Fukushima right after the accident and began the in situ research on a voluntary basis. The contaminated area of agriculture means the environment itself and considering the various kinds of contaminated places or things, there is no other group consisting of so many kinds of experts working together systematically to determine the real-time situation of radio-contamination.

It takes a long time to determine where radioactive cesium will accumulate and how they are moving. For example, when a radioactive crop is found and we want to further investigate the change in the accumulation manner, we must grow the plants in the same culture field and wait for the next year. That means that we can obtain one result per year.

In these 10 years, we showed the features of the radio-contamination caused by fallout, which were completely different from those caused by heavy metals. In particular, the chemical behavior of fallout, carrier-free radionuclides, is not well known; therefore, chemical decontamination is very difficult to perform. However, the carrier-free nuclides did not move and stayed at the original place when they were adsorbed. The strong adsorption of radioactive cesium on soil resulted in less contamination of the products and fewer health effects on humans. When we know these features of the fallout well, not only the method of radiation protection but also the decontamination method might be changed.

The Fukushima Nuclear Power Plant accident was the first accident to occur in the monsoon zone on Earth. Although the nuclear power plant accident must not occur again, when it occurs in the template region, the results we presented in the series of four books will provide the proper way to understand the behavior of the fallout. Since they are all open access, we wish the books will be read by as many people as possible.

Tokyo, Japan   Tomoko M. Nakanishi

Acknowledgement

We would like to thank all the people who have been continuing the effect of Fukushima Daiichi Nuclear accident for more than 10 years. Though we are in the group of 50–60 researchers, in these years some of them are moved to the other universities or institutes. However, most of them are still performing this research and carrying out the actual experiments. We also thank the people live at Fukushima area, especially Mr. Kinichi Okubo and the people in the group, named Fukushima-Saiseinokai, and the other people who supported and cooperated the work with us. Without their help, it was impossible for us to continue the measurement or experiment at Fukushima.

Lastly, we would like to thank Mr. Kazumasa Ikushima, the president of Musashi Engineering, Inc., who kindly supported us to publish this book as an open access.

Contents

Chapter 1
An Overview of Our Research

Tomoko M. Nakanishi

1.1 Introduction

In these 10 years, the radioactive cesium distribution and movement were surveyed. The results we found at Fukushima were presented in the paper and the former 3 books of this series (Nakanishi 2018; Nakanishi and Tanoi 2013, 2016; Nakanishi et al. 2019). The most important result was that the fallout was adsorbed firmly on the surface of any materials exposed to the air at the time of the accident and then hardly moved or even washed. This kind of nature is different from the well-known chemical behavior of the cesium we experience in the laboratory. This element behavior shown by a very small number of elements is known as the radio-colloidal nature, which is mainly discussed in radiochemistry. We could detect radioactive cesium, because it emits radiation, and in principle, we can detect only one element by radiation measurement. Such a high sensitivity to detect one atom could not be accomplished by the other analytical tools. Therefore, the method of decontamination was rather simple, just removing the surface of anything contaminated, since the radioactive cesium does not move. In the case of soil, radioactive cesium adsorption was featured on a small soil component, clay. The adsorption of the fallout was very strong so that most of the plants grown in soil absorbed a very limited amount of radioactive cesium from the soil, which dramatically reduced the radio-contamination of the agricultural products grown in the field.

We have been continuing the research to determine the behavior of radioactive cesium at contaminated sites for more than 10 years, and now, through radioactive analysis, the data allowed us to estimate the future behavior of radioactive cesium. To get an overall picture of the degree of contamination, we studied the radioactive cesium effects on soil, water, crops, animals, mountain areas, etc., and a

T. M. Nakanishi (✉)
Graduate School of Agricultural and Life Sciences, The University of Tokyo, Tokyo, Japan
e-mail: tmn@g.ecc.u-tokyo.ac.jp

© The Author(s) 2023
T. M. Nakanishi, K. Tanoi (eds.), *Agricultural Implications of Fukushima Nuclear Accident (IV)*, https://doi.org/10.1007/978-981-19-9361-9_1

1

comprehensive explanation on radio-contaminations is described here; this comprises an extension of our research we already presented in the former books of this series.

1.1.1 Soil

Since the accident was in late winter, there were hardly any plants growing in the field, except for wheat. In the agricultural field, everything exposed to the air was contaminated, including the leaves of wheat, and the main contaminated place was the soil surface. It was very difficult to remove radioactive cesium from the soil, since it was firmly adsorbed on the soil and was hardly moved even under hard rain.

It was estimated that radioactive cesium will go downward with time, since global fallout, which occurred during the 1960s due to the atomic bomb test, and is still able to be measured in many places throughout Japan. When the radioactivity of this fallout was measured, it was found that they were found at a certain depth of the soil, indicating that they moved downward and could not be detected at the surface. In the case of the Fukushima accident, to measure the downward speed of radioactive cesium, pipes were prepared and buried in a straight from the surface of the soil. Periodically, the counter was moved in the pipe from the upper part to the bottom to measure the radioactivity profile in the soil with depth. The movement of radioactive cesium during the first few weeks was relatively fast, but the movement slowed gradually with time, and in these years, the movement of radioactive cesium was approximately 1–2 mm/year.

In the case of global fallout, the measurement was performed in a place remote from the Fukushima area, where fallout by the Fukushima Nuclear Accident was not found. The moving distance of radioactive cesium from the surface, after approximately 60 years, was approximately 8 cm, indicating that the moving speed of radioactive cesium from the fallout of Fukushima Nuclear Accident was roughly the same.

Then, the soil particle adsorbing a high amount of radiocesium was identified. Only in the fine clay, called weathered biotite, was highly accumulated the radiocesium. Other components of the soil, especially larger than the fine clay, radioactive cesium was not found. Then, the method of adsorption within the clay was studied. The absorption in the fine clay was very tight, and the distribution of radioactive cesium was found to be uniform within the clay. Sometimes, the edge of the clay has been discussed to confine the radiocesium; however, such an edge was difficult to observe.

Since contaminated soil was only at the surface, within approximately 5 cm from the surface, removing the soil surface was the effective method of decontamination. The soil surface was removed from many farming lands using heavy machinery. However, machines used on the fields pressed the soil so that the air layer in the soil was lost. To grow the plants again in contaminated fields, first, air must be introduced, and the uniform growth of the plants must be resumed. However, these activities are difficult to perform in many areas, because the farmers are getting too old for these works.

The features of the soil contamination found were as follows

- The soil surface, less than 5 cm from the surface, was highly contaminated
- Hardly moved downward with hard rain
    The moving speed was not related to the amount of rain
- The speed of the downward movement of radioactive cesium was approximately 1–2 mm/year
    This speed was found to be similar to that of the global fallout during the 1960s, which can still be measured in Japan
- Most of the radioactive cesium was adsorbed tightly in fine clay particles
    The distribution of the radioactive cesium in the clay was homogeneous
    Particles larger than fine clay were not contaminated
- After the removal of the soil surface, there are other problems to restart farming, such as the introduction of the air layer in the soil and the resumption of homogenous plant growth

1.1.2 Water

There are two forms of radioactive cesium in water, dissolved or suspended. The water in the rivers investigated contained only suspended radioactive cesium. The radioactive cesium was contained firmly in the fine clay and moved with water in a suspended form. Therefore, the suspended cesium was easily separated by simple filtration.

When river water, as well as soils under water, was sampled periodically along the river, it was found that when the water moving speed was fast, most of the suspended cesium was moved with water in the river. However, when the moving speed slowed, for example, at the inside curve of the meandering river, suspended clay was sedimented.

To determine whether biological accumulation occurs by any plants, insects, or animals, many samples along the food chain were collected and measured. The result was that there was no specific living thing found that accumulated radioactive cesium.

Then, the next investigation was to determine how much radioactive cesium moves with water or how much radioactive clay comes out from the mountain. Several devices were placed throughout the mountain so that the water flowing speed and the actual water amount moving on the soil as well as on the trees were analyzed. When there was a large amount of rain, the amount of flowing water increased, and the radioactivity of the water collected was high at the beginning of the rain. However, soon after the first flow, there is a high amount of water coming out from the soil, since mountains store a high amount of water in the soil. Therefore, the radioactivity found in the initial water was soon diluted with the reserved water flowing out from the mountain.

In the case of water reservoirs, unexpected results were found. When the soil surface under the water, at the bottom of the pond, was measured, highly contaminated soil was found. To determine the origin of the radioactive cesium, tracing the route of the river and using the land were investigated. Only in the upper part of the land of this water reservoir containing highly radioactive soils was decontamination activity a vigorously conducted area. This means that the radioactive cesium removed by hard washing flowed out from the residential part and flowed into the pond. Additionally, it was found that the cesium adsorbed on concrete or paved surfaces could be removed to some extent by hard washing, causing secondary contamination of the pond. Once it flew into the pond, the radioactive cesium was adsorbed firmly on the soil surface at the bottom of the pond.

The feature of water contamination found can be summarized as follows

- Dissolved radioactive cesium was rarely found.
- Radioactive cesium adsorbed on fine clay was suspended in water.
- Clay with radioactive cesium was separated from water by simple filtration.
- With a decrease in water flow, some of the suspended radioactive cesium was sedimented.
- Radioactive soil that comes out from mountains by water flow was approximately 1/1000 of the fallout.
- There was no living thing found that specifically accumulated radioactive cesium.
- Vigorous decontamination activities of concrete or paved roads partially remove radiocesium, and the decontamination water flows into the water reservoir.

1.1.3 Mountain

Since the priority of decontamination was focused on human activity-related areas, including agricultural lands, most of the mountain area was left behind. The other reason was that the mountain area is large, approximately 70% of Fukushima prefecture, so it was difficult to cover the whole area of the mountain to decontaminate.

What had happened to the mountain was like the farming land mentioned above, everything exposed to the air was contaminated. Most of the mountains were contaminated by coniferous forest; therefore, the needle-like leaves and the bark of the trees were highly contaminated. The degree of contamination was higher with respect to height. In the case of branches, the upper sphere accepted more radiocesium compared to that of the lower part, facing the soil.

To observe the inner contamination of the tree, cedar and pine trees were downed, and the wood disks were prepared from upper part to bottom. Then, it was found that the radiocesium adsorbed on the bark was partially moved into the inner part of the trunk. The amount of radiocesium moved inside was higher at higher positions of the trunk and higher in cedar trees than in pine trees. There was a freshly downed cedar tree, and the radioactivity inside the tree was also measured. Interestingly, even in the downed tree, the radiocesium was moving inside the tree.

After several years, contaminated conifer leaves or part of the bark of small branches fell to the ground and increased the radioactivity on the ground. Then, the radioactive cesium accumulated in these tissues was transferred to the soil surface during the process of decomposition by microorganisms. Therefore, in most forests, radioactive cesium is now moving toward the soil surface.

Mushrooms have a problem, since they collect radiocesium during growth, and the radiocesium circulates around the small area where mushrooms grow. It was so interesting that many mushrooms growing in the area where fallout from Fukushima cannot have any influence are still collecting or maintaining the radiocesium they collected at the time of global fallout, approximately 60 years ago.

The features of the mountains found were as follows

- Leaves and bark of the tree was highly contaminated.
- The degree of contamination was higher with respect to the height.
- Part of the radioactive cesium was moved inside the tree.
- The amount of radiocesium inside the tree was higher in cedar trees than in pine trees.
- Even the cut-down trees contaminated with fallout, radiocesium moved inside the tree.

  - After several years, the contaminated needle-like leaves in coniferous trees or part of the bark or branches fell down to the ground.
  - The tissue that fell from the trees was decomposed by microorganisms, and the radiocesium adsorbed on the tissue was transferred to the soil.

- Gradually, the soil surface in the mountains has been accumulating high amounts of radiocesium.

1.1.4 Others

Animals grown outside are also contaminated. However, when noncontaminated food was supplied, most of the radioactive cesium was metabolized to the outside of the body after several months. Although the half-life of cesium-137 is 30 years, its biological half-life is estimated to be within 100 days. In the case of the meadow, a trial to propose a new agricultural system has been introduced. For example, how to establish the circulation of metabolites using contaminated compost. Wild birds were found to be contaminated; however, the birds were found to be decontaminated the following year through the renewal of feathers. The contamination of each organ in wild boar was also investigated.

The most effective way to reduce the uptake of radiocesium to the plant is to supply potassium to the field. Although potassium is an essential element for plants to grow, the amount of potassium for pasture should be kept low, since more potassium causes disease in animals.

1.2   Conclusion

The essential point is that the fallout does not move, which makes a large difference from that of the contamination caused by heavy metals we experienced before. Heavy metals are dissolved in water and accumulate in plants, fish, or animals, which causes serious health effects on people who eat them. However, in the case of fallout, most of them did not dissolve in water, therefore, was not accumulated highly in living things; therefore, when supplied as foods, they did not cause any serious health effect in human beings.

As we pointed out, we could separate the contaminated soil from that of the matrix soil, showing a clue to decontaminate the soil. One of the methods is, as we have shown, the introduction of water to the field to separate the fine particles and leave most of the matrix there. What we must consider is the reservation of the soil itself, which is an indispensable resource to grow serials or vegetables in the field. Since it takes so long a time to create soil in nature, we always must pay close attention to it.

References

Nakanishi TM (2018) Agricultural aspects of radiocontamination induced by the Fukushima nuclear accident—a survey of studies of the University of Tokyo Agricultural Department (2011–2016). Proc Jpn Acad Ser B 94:20–34
Nakanishi TM, Tanoi K (eds) (2013) Agricultural implications of the Fukushima nuclear accident. Springer, Tokyo
Nakanishi TM, Tanoi K (eds) (2016) Agricultural implications of the Fukushima nuclear accident. The first 3 years. Springer, Tokyo
Nakanishi TM, O'Brien M, Tanoi K (eds) (2019) Agricultural implications of the Fukushima nuclear accident (III). After 7 years. Springer, Tokyo

Chapter 2
Recovery of Food Production from Radioactive Contamination Caused by the Fukushima Nuclear Accident

Naoto Nihei

2.1 Post-Fukushima Daiichi Nuclear Power Plant Accident Safety Measures for Agricultural Products Produced in Fukushima Prefecture

Following the Great East Japan Earthquake on March 11, 2011, the accident at Tokyo Electric Power Company's Fukushima Daiichi Nuclear Power Plant resulted in the spread of contamination by radioactive materials (mainly radioactive cesium) over a wide area of eastern Japan, centered on Fukushima Prefecture. When the concentration of radioactive cesium in the soil (surface layer 15 cm) was 5000 Bq/kg or less, inversion tillage (deep plowing) was used, and when the concentration was more than that, topsoil stripping at a thickness of 5 cm or more was used. Decontamination work was completed in all areas except the difficult-to-return zone by the end of March 2017. Decontamination had been completed for approximately 22,000 housing sites, 8400 ha of farmland, 5800 ha of forests, and 1400 ha of roads. After decontamination, it is recommended that agricultural fields maintain a high exchangeable potassium content in the soil to prevent radioactive cesium absorption (the standard for rice cultivation is 25 mg/100 g or more), and potassium has been cultivated by adding potassium fertilizer to the usual potassium application. Furthermore, the epidermis of fruit trees was washed with water and peeled off to remove radioactive cesium.

To ensure the safety of agricultural products grown on farmland after decontamination, emergency environmental radiation monitoring (hereinafter referred to as "monitoring inspection") based on the Act on Special Measures Concerning Nuclear Emergency Preparedness has been conducted since immediately after the accident.

N. Nihei (✉)
Faculty of Food and Agricultural Sciences, Fukushima University, Fukushima City, Fukushima, Japan
e-mail: nihei@agri.fukushima-u.ac.jp

T. M. Nakanishi, K. Tanoi (eds.), *Agricultural Implications of Fukushima Nuclear Accident (IV)*, https://doi.org/10.1007/978-981-19-9361-9_2

The inspection focuses on agricultural, forestry, and fishery items that are offered for sale in Fukushima Prefecture, and these products are inspected before being sent. Under the Food Sanitation Law, the Ministry of Health, Labor, and Welfare set the interim regulating values for radioactive materials as 2000 Bq/kg for radioactive cesium and 500 Bq/kg for radioactive cesium on March 17, 2011. Furthermore, since April 2012, the standard value for the concentration of radioactive cesium in general foods has been set at 100 Bq/kg. If the concentration of radioactive cesium exceeds the standard value as a consequence of monitoring inspections, shipments are stopped by each municipality and will not be distributed to the market. During the 10 years following the accident until March 2021, Fukushima Prefecture conducted monitoring inspections on about 233,000 samples (excluding rice, grass, and other plants) in around 500 products, and is continuing to do so while narrowing down the items to be inspected (Fukushima Prefecture Monitoring of Agricultural, Forestry, Fishery and Processed Food Products Information HP 2021).

Rice is the most valuable agricultural commodity in Fukushima Prefecture and is the staple food of the Japanese people. Resultantly, in 2012, instead of a sampling inspection like the monitoring inspection, a more complete inspection of all rice produced in Fukushima Prefecture (estimated to be over 360,000 tons per year) was done [hence referred to as "inspection of all rice bags" (Fukushima No Megumi Safety Council HP 2021)]. A measuring equipment (generally known as a belt-conveyor type survey meter) was created to conduct the examination of all rice bags and determine if the concentration of radioactive cesium in a bag of rice (30 kg) is below the standard value in a few tens of seconds. Fukushima Prefecture installed approximately 200 belt-conveyor type survey meters and conducted inspections in conjunction with growers' shipments. At the end of the inspection, an individual identification number is added to each bag of rice along with a sticker stating that it has been inspected, and the individual results can be viewed on the website.

2.2 Current Status of Agricultural Products from Fukushima Prefecture

Figure 2.1 shows the monitoring results of agricultural (crop, vegetables, fruit, etc.), livestock, forestry (wild vegetable and mushroom), and fishery (saltwater fish and freshwater fish) products from 2011, and the dotted line indicates the current standard value of 100 Bq/kg. In FY2011, immediately after the nuclear power plant accident, the percentages for grains (excluding rice), vegetables, and fruits were 7.2%, 4.7%, and 11.2%, respectively, but have declined dramatically after FY2012. Since FY2016, FY2013, and FY2018, respectively, no samples exceeding 100 Bq/kg have been detected in cereals (excluding rice), vegetables, and fruits. The high percentage of samples exceeding 100 Bq/kg in FY2011 was due to the direct fallout (direct contamination) of radioactive cesium emitted from the nuclear power plant on agricultural products (wheat, spinach, komatsuna, etc.) grown in the field.

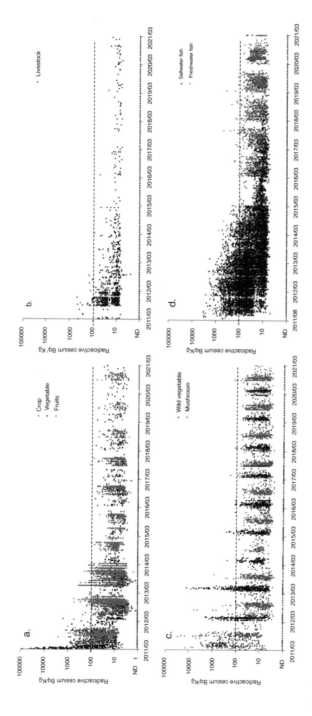

Fig. 2.1 The monitoring results of agricultural (**a**), livestock (**b**), forestry (**c**), and fishery (**d**) products from 2011

Furthermore, many of the few samples of produce that exceeded 100 Bq/kg since 2012 were found to be primarily due to the use of agricultural equipment contaminated by radioactive materials during cultivation, rather than absorption and transfer from soil to crop. Therefore, administrative guidance was also provided in Fukushima Prefecture to be very careful in the use of agricultural materials and machinery. Since 2012, roughly ten million rice bags (30 kg) have been inspected annually as part of a comprehensive inspection of all rice bags, separately from monitoring inspection. As a result, 71 bags of rice exceeding 100 Bq/kg were detected in 2011, 28 bags in 2012, and 2 bags in 2013, and no rice exceeding 100 Bq/kg has been detected since 2014. From 2020, all bags are no longer inspected using a belt-conveyor type survey meter, except for 12 of the 59 municipalities in Fukushima Prefecture that have been ordered to evacuate due to the nuclear power plant accident.

In FY2011, the percentage of meat that exceeded 100 Bq/kg in FY2011 was 0.8%. The reason why radioactive materials were detected in beef and pork was thought to be the feeding of feed contaminated with radioactive materials, so farmers were warned not to feed rice straw and grass that was outside at the time of the accident to their cattle and pigs, and administrative guidance was provided to prevent contamination of meat from spreading, and since FY2012, no radioactive materials have been detected.

For saltwater and freshwater fish, the percentages exceeding 100 Bq/kg were 24.5% and 30.8%, respectively, in FY2011, 12.7% and 12.9% in FY2012, 2.3% and 8.0% in FY2013, 0.6% and 2.8% in FY2014. Since FY2015, few saltwater fishes have been detected above 100 Bq/kg (one point exceeded 100 Bq/kg in January 2019), while 0.3% of samples of freshwater fishes were still above 100 Bq/kg in FY2019. Fishery products are also steadily declining, but some marine products (rivers and inland waters) continue to be detected. The reasons for this include contamination of food (fallen leaves and insects) and the fact that freshwater fish do not discharge much radioactive cesium that they have taken in in an attempt to maintain the salt concentration in their bodies.

The percentages exceeding 100 Bq/kg for wild vegetables and mushrooms were 35.5% and 19.4% in FY2011, 18.2% and 12.5% in FY2012, 11.6% and 4.5% in FY2013, 3.4% and 4.4% in FY2014, respectively. Since FY2015, mushrooms have not been detected above 100 Bq/kg, whereas wild vegetables were detected until FY2018. Differences were observed among wild vegetables, with a tendency for high concentrations in young leaf-eating wild vegetables such as Koshiabra (*Eleutherococcus sciadophylloides*). The concentration of radioactive cesium in wild vegetables and mushrooms has also decreased markedly over the years. However, the reason why they are higher than other items is because the mountains where they grow are undecontaminated and potash fertilizer is not applied to suppress the absorption of radioactive cesium.

2.3   Future Prospects

Some farmers in Fukushima Prefecture were unable to farm due to the evacuation zone or were apprehensive about selling their produce, and the utilization rate of arable land fell dramatically following the nuclear power plant accident (FY2011: National average 92%, Fukushima prefecture 75%). Ten years have passed since the nuclear accident, and evacuation orders have been lifted in many areas, but the arable land utilization rate in FY2018 is still at the same level as it was immediately after the nuclear accident (Fukushima Prefecture 2021). One of the factors behind the lack of a return to arable land use is that the rate of return has not increased even after the evacuation order has been lifted. The reasons why the return rate has not increased is that, in addition to the fact that people have established their lives in the evacuation sites after the nuclear accident, life in the satoyama, where people live in communities close to forest, has also disintegrated. Many of the areas where evacuation orders were issued are surrounded by mountains, and for people living in the mountainous areas, growing rice in the paddies at home and gathering wild vegetables and mushrooms in the neighborhood were part of their daily life. Even after decontamination is complete and rice can be grown in the paddies, some people are hesitant to return to their homes, because their enjoyment will be cut in half if they cannot eat wild vegetables and mushrooms. Although the economic evaluation of collecting wild vegetables and mushrooms for personal consumption may be low, it is a significant concern for the locals, and the restoration of satoyama life is one of the issues that will be addressed in the future.

Furthermore, while few samples exceeding the standard values have been detected in monitoring inspections, but there are still concerns regarding radioactive contamination of agricultural products from Fukushima Prefecture. Furthermore, unlike before the nuclear accident, rice produced in the areas near the Fukushima Daiichi Nuclear Power Plant (Hamadori and Nakadori) is increasingly employed for commercial purposes, and the price of rice has remained low. To dispel such rumors, the results of monitoring inspections have been publicized, and aggressive marketing activities have been conducted in the Tokyo metropolitan area.

Regarding cultivation, potassium fertilization, which has been implemented as a measure to control the absorption of radioactive cesium, is gradually being phased out. However, considering the half-life of radioactive cesium, it will take a long time for all cesium in the agricultural environment to disappear; therefore, regular monitoring is required to maintain safety. The livestock industry in Fukushima Prefecture is also thriving, but because grass and other feed containing high potassium content can cause grass tetany in cattle, it is not possible to rely solely on potassium-enriched fertilizers, and the prefecture is struggling with countermeasures. Research on rice types that are less likely to absorb cesium has made progress (Nieves-Cordones et al. 2017; Rai et al. 2017), which is expected to revitalize agriculture, and it is hoped that this will be applied to pasture grasses as well.

Fukushima Prefecture is Japan's third largest prefecture in terms of geographical area, and its agriculture, forestry, and fisheries industries are rich in regional

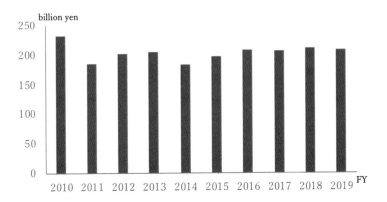

Fig. 2.2 Fukushima Prefecture's agricultural product value from 2010 to 2019

characteristics, ranging from regions with mild winters and long hours of sunshine to inland regions with large daily temperature differences. The prefecture is rich in natural resources and is a leading producer of rice and other agricultural products such as cucumbers, tomatoes, pea pods, *Asparagus*, and peaches. Fukushima Prefecture's agricultural product value decreased greatly after the nuclear accident (2011: 185.1 billion yen) compared to prenuclear accident (2010: 233 billion yen) but rebounded to 208.9 billion yen in 2019 (Fig. 2.2). In the evacuation zone, specific reconstruction and revitalization centers have been established, and efforts to regenerate agriculture have begun, albeit gradually. Although some areas have aged rapidly due to the nuclear accident, farmers in Fukushima Prefecture are struggling to engage in agriculture in new ways, such as opening vineyards with wine breweries, forming sweet potato and flower production areas, improving efficiency through smart agriculture, and exporting agricultural products from Fukushima Prefecture. Farmers are concerned about the loss of production on soil that has lost its crop layer, as well as whether their products can be marketed. Continued monitoring inspections and promoting understanding of produce based on correct information are necessary for agricultural recovery in the affected areas.

References

Fukushima No Megumi Safety Council HP (2021). https://fukumegu.org/ok/contentsV2/

Fukushima Prefecture (2021) Current status of agriculture, forestry and fisheries in Fukushima Prefecture. https://www.pref.fukushima.lg.jp/sec/36005b/norinkikaku2.html

Fukushima Prefecture Monitoring of Agricultural, Forestry, Fishery and Processed Food Products Information HP (2021). https://www.new-fukushima.jp

Nieves-Cordones M, Mohamed S, Tanoi K, Kobayashi NI, Takagi K, Vernet A, Very A (2017) Production of low-Cs^+ rice plants by inactivation of the K^+ transporter OsHAK1 with the CRISPR-Cas system. Plant J 92(1):43–56

Rai H, Yokoyama S, Satoh-Nagasawa N, Furukawa J, Nomi T, Ito Y, Hattori H (2017) Cesium uptake by rice roots largely depends upon a single gene, HAK1, which encodes a potassium transporter. Plant Cell Physiol 58(9):1486–1493

Chapter 3
Annual Reduction of Transfer Factors of Radiocesium from Soil to Rice Cultivated in a KCl Fertilized and Straw Plowed-in Paddy Field from 2015 to 2021

Ichio Ii

3.1 Introduction

Rice is the most important crop in Japan. The Tokyo Electric Power Company's Fukushima Daiichi Nuclear Power Plant (TEPCO-FDNPP) accident in March 2011 caused radioactive materials to spread out and down onto farmlands around the Plant. In the Iitate village more than 30 km far from the Plant in Fukushima Prefecture, people were evacuated from June 2011 to March 2017, except Nagadoro Ward that still needs to be evacuated. In the village, we performed consecutive field trials of rice cultivation and monitored radiocesium contamination in rice from 2012 to judge how we can cultivate rice in the village again, without fear of radioactive contamination. The early year results (2012–2013) (Ii et al. 2015; Ii and Tanoi 2016) showed KCl fertilization can reduce the transfer factor (TF) of radiocesium in brown rice from soil to 0.003–0.004 in 2012 when the exchangeable K content in soil was higher than 20 mg/100 g (as K_2O). From 2014, all bags of brown rice harvested at Sasu test field have passed the Fukushima Prefecture inspection (below the detection level of 25 Bq/kg with screening measurement). However, consumers prefer food with less radiocesium contamination and farmers usually recycle straw for the following year rice plantation. Then we continued to cultivate rice at KCl fertilized and straw plowed-in paddy field and monitor the radiocesium concentration of brown rice and straw harvested. Similar results of importance of the exchangeable K content were reported by Fujimura and Eguchi (2016) at five paddy fields in

I. Ii (✉)

NPO Resurrection of Fukushima, Sasu Aza Nameri, Iitate, Soma-Gun, Fukushima Prefecture, Japan

Circle Madei, Graduate School of Agricultural and Life Sciences, The University of Tokyo, Tokyo, Japan

© The Author(s) 2023
T. M. Nakanishi, K. Tanoi (eds.), *Agricultural Implications of Fukushima Nuclear Accident (IV)*, https://doi.org/10.1007/978-981-19-9361-9_3

Fukushima Prefecture in 2012–2014. Further Yamamura et al. (2018) proposed a statistical model for estimating the radiocesium TF from soil to brown rice using the soil exchangeable K content, based on the field data obtained from 2012 to 2015. From their model, they also estimated the year factor and a yearly decline rate of 17%.

On the other hand, Roig et al. (2007) tested the conditions of long-term radiocesium trapping (ageing) by clay soil from the Chernobyl area, and Yamashita et al. (2016) reported yearly decrease in the TF of radiocesium from soil to grasses from 2012 to 2015 on a pasture in Iwate Prefecture. Tsukada (2014) reported that the ratio of radiocesium in the exchangeable fraction decreased from 12.8% in April of 2012 to 6.9% in October of 2013 in a paddy field in Fukushima. Further Fujimura and Eguchi (2016) showed 40% and 30% decrease in TF in brown rice and inedible rice part from 2011 to 2012, by employing pot experiments using soil samples from Fukushima, suggesting an irreversible sorption of ^{137}Cs to clay minerals. Tagami et al. (2018) also showed the geometric mean of ^{137}Cs TF of brown rice to soil from paddy fields without additional K decreased from 0.012 in 2011 to 0.0035 in 2013. Wakabayashi et al. (2020) reported ^{137}Cs ageing assessed by the ratios of exchangeable ^{137}Cs/^{133}Cs in a field study of a rice paddy in allophanic Andosol in Tsukuba from 2011 to 2015. All the data reported are from those in 4 years after the accident. Here, we report yearly decrease in the TF of ^{137}Cs from soil to brown rice and straw from 2015 to 2019, based on Ii et al. (2021) and to bottom levels in 2019 to 2021. The decrease from 2015 to 2019 is suggested to be mainly due to ageing effect.

3.2 Materials and Methods

3.2.1 Test Field, Rice Cultivation, and Sampling

Rice cultivation is performed at the same field at Sasu test field (N37°44′, E140°43′) as reported (Ii et al. 2015, 2021; Ii and Tanoi 2016) (Fig. 3.1).

The soil is classified as an Andosol (D2z1) by Japanese soil inventory system (http://soil-inventory.dc.affrc.go.jp). The field was partially decontaminated in April 2012 by shallow irrigation and the muddy water swept out from North-East to South-West as reported (Ii et al. 2015, 2021; Ii and Tanoi 2016). The soil radiocesium concentration was 2000–6000 Bq/kg of dry weight. Further decontamination was not performed thereafter. In 2012 and 2013, the straw harvested was not plowed in, for fear of possible radiocesium contamination. In 2014 and after, all the straw harvested was plowed in the same paddy field as fertilizer for the following cultivation. Rice seedlings of Akitakomachi (2012–2013), Hitomebore (2014–2017, 2020–2021), Koshihikari (2018–2019) were planted in the fields. Rice radiocesium absorption is not significantly different among these cultivars (Ono et al. 2013). Table 3.1 summarizes rice cultivation procedures including fertilizers used from 2014 to 2021.

Rice planting was performed in May or early June. In 2014, a basal fertilizer (18N–15P–15K; weight % as N, P_2O_5, and K_2O, 50 kg/10a) was employed and K

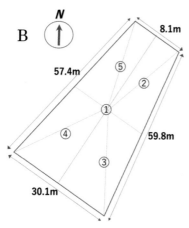

Fig. 3.1 The test field of Sasu and the 5 sampling points (**a**) the picture of rice harvest with hand, in October 6 of 2019, the next day of the sampling (sampling points M1–5 shown). (**b**) Diagram of the sampling points (1–5). [Figures from Ii et al. (2021)]

fertilizer was added as KCl (60% as K_2O:20 kg/10a) to south-east half area (K test field) and mixed with the plowed soil before planting. In 2015, the basal fertilizer (18N–15P–15K, 50 kg/10a) and KCl (20 kg/10a) were added to whole field. In 2016, the same as in 2015, except an additional fertilizer of NK (17N:17K) 2–3 kg/10a was added in July. In 2017, the same as in 2016, except the addition of Ca silicate (50 kg/10a) as a soil improver and a basal fertilizer (20N–15P–14K, 50 kg/10a) was used. In 2018 and 2019, the same as in 2017, except for no soil improver added and the amount of additional NK fertilizer (17N–17K) of 4–5 kg/10a in July. In 2020 and 2021, the same as in 2019 except the amount of additional NK fertilizer of 2–3 kg/10a. In 2021, Koshihikari was also planted in a small part just for a comparison purpose.

After sampling by hand, harvest was performed in October and the remnant of straw was plowed in the paddy field. In September 10 in 2015, typhoon 18 caused

Table 3.1 Rice cultivation procedures and sampling date (2014–2021)

Year	Species	Basal fertilizer	Additional fertilizer	Straw plowed in	Sampling date	Note
2014	**Hitomebore**	Fertilizer or Hitomebore (18–15–15) 50 kg/10a; KCl 20 kg/10a (KCl addition half area only)	No	Yes	Oct. 5	
2015	**Hitomebore**	Fertilizer or Hitomebore (18–15–15) 50 kg/10a; KCl 20 kg/10a	No	Yes	Oct. 3	Sep. 10, Typhoon 18 caused brook over-flow with soil and sand to the test field
2016	**Hitomebore**	Fertilizer for Hitomebore (18–15–15) 50 kg/10a; KCl 20 kg/10a	NK fertilizer (17–0–17) 2–3 kg/10a	Yes	Oct. 1	
2017	**Hitomebore**	Fertilizer for Hitomebore (20–15–14) 50 kg/10a; KCl 20 kg/10a (Ca Silicate 50 kg/10a as soil improver just before addition of fertilizer)	NK fertilizer (17–0–17) 2–3 kg/10a	Yes	Oct. 9	
2018	**Koshihikari**	Fertilizer for Hitomebore (20–15–14) 50 kg/10a; KCl 20 kg/10a	NK fertilizer (17–0–17) 4–5 kg/10a	Yes	Sep. 29	
2019	**Koshihikari**	Fertilizer for Hitomebore (20–15–14) 50 kg/10a; KCl 20 kg/10a	NK fertilizer (17–0–17) 4–5 kg/10a	Yes	Oct. 5	Oct. 11–12, Typhoon 19 caused brook over-flow with soil and sand to the test field
2020	**Hitomebore**	Fertilizer for Hitomebore (20–15–14) 50 kg/10a; KCl 20 kg/10a	NK fertilizer (17–0–17) 2–3 kg/10a	Yes	Oct. 2	

(continued)

Table 3.1 (continued)

Year	Species	Basal fertilizer	Additional fertilizer	Straw plowed in	Sampling date	Note
2021	**Hitomebore Koshihikari**	Fertilizer for Hitomebore (20–15–14) 50 kg/10a; KCl 20 kg/10a	NK fertilizer (17–0–17) 2–3 kg/10a	Yes	Oct. 2	

Data for 2014 to 2019 from Ii et al. (2021)

the brook overflow at the south-west side, and the soil and sand covered part of the south-west side but not to the sampling points in the field and soil and rice sampling was performed at 23 days after the flood. In 2015–2021, soil and rice sampling were performed between late in September and early in October at the 5 points (M①, M②, M③, M④, M⑤) of the test field, as shown in Fig. 3.1. The soil of 0–15 cm at each point was taken into a plastic bag with a long scoop, and 10–15 sheaves of rice plant were cut and collected. The bundles of rice plant were threshed with a foot-driven thresher to give one unhulled rice sample for each test point at Sasu (Fig. 3.2).

The soil sample and the unhulled rice sample were sent to "Circle Madei" (Ii and Tanoi 2016), a volunteer employee and student group at Tokyo University and three aliquots of co. 20 mL of soil in vials were prepared from each soil in a bag for the radioisotope measurement (Fig. 3.3). The unhulled rice sample was kept for a week or more in the room, and then brown rice sample was prepared with a hulling machine. The straw sample was dried naturally in a greenhouse for a week or more. Then the straw sample for radioisotope measurement was prepared by cutting the straw part by 1 cm, sent to "Circle Madei," and further dried in the room for a week or more.

3.2.2 Measurement of Radiocesium and Exchangeable Cations

The ^{137}Cs in brown rice and straw were measured by a Ge semiconductor detector (GEM and GMX type; Seiko EG&G) for 1–24 h in 100 mL or 250 mL containers. The ^{137}Cs value above each detection level of 2σ was adopted. The water content of brown rice and straw samples were around 10%. The values were adjusted as 10% water content. Further to calculate TF, the values are adjusted to the date of measurement of the soil, using a half-life of 30.08 year. The soil samples were measured by a NaI (Tl) scintillation counter (2480WIZARD2Autoγcounter:Perkin Elmer) in 20 mL vials (Nobori et al. 2013) within a week after the sampling date. The value of the soil was corrected per dry weight by measuring the soil weight after drying the soil at 60 °C for more than 6 days. The exchangeable cations in the soil extracts were analyzed using ICP-OES (Optima 7300DV) (Ii et al. 2015). The dry

Fig. 3.2 Photos of the sampling of rice and soil (left), and the rice bundles were threshed with a foot driven thresher (right)

 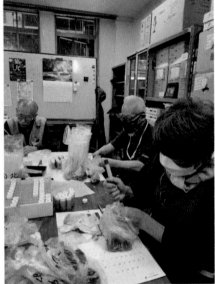

Fig. 3.3 Photo of recently renewed "Circle Madei" room, the members preparing soil samples for measurement of radiocesium

soil was crushed and sieved with 2 mm sieve, and 4 or 10 g of the sieved dry soil was mixed with 10 vol. of 1 M ammonium acetate solution at room temperature for more than 1 h and the supernatant was obtained by centrifugation and the filtrate with 0.2 μm filter was used for the analysis. For the analysis of exchangeable ^{137}Cs, 10 g of the sieved dry soil and 100 mL of ammonium acetate was used, and 80 mL of

0.2 μm filtrate was used for the measurement by a Ge semiconductor detector. The data was divided by ^{137}Cs value of 8 g of the dry soil measured by the NaI scintillation counter and adjusted to the Ge measurement date, using the half-life. For the measurement, the dry soil was prepared at 60 °C for 7 days and stored at room temperature (15– 25 °C) for 15 months in case of 2017 soil and for 3 months in case of 2018 and 2019 soils. The 2017 soil and 2018 soil were measured simultaneously. To compare the soils in 2019, 2020 and 2021, 2019 soil stored for 15 months, 2020 soil for 3 months and 2021 soil was stored for 5 months. The 2019 and 2020 soils were measured simultaneously.

3.3 Results and Discussion

3.3.1 Yearly Change of ^{137}Cs in Brown Rice, Straw and Paddy Soil

Figures 3.4 and 3.5 show the yearly change of ^{137}Cs in brown rice (a) and straw (b) from 2015 to 2021 at each sampling point. Though some data of brown rice below the detection level were not presented in Fig. 3.4a and one missing data in Fig. 3.5b, annual decrease of ^{137}Cs in brown rice and in straw are obvious, though variance is observed. The variance may be due to possible contamination with dirt and sampling variance.

Figure 3.6 shows yearly results of the ^{137}Cs concentration of paddy soil at each sampling point (M①, M②, M③, M④ and M⑤) from 2015 to 2021 as shown in

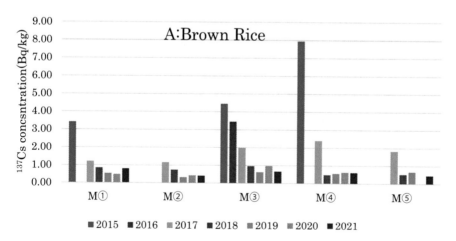

Fig. 3.4 The yearly change of ^{137}Cs in brown rice (**a**) at the sampling points (2015–2021). The values are adjusted to the date of measurement of the soil within a week after the rice sampling. Data are not presented for M2 in 2016, M5 in 2015, 2016, and 2020, because they were below the detection level (2σ). [Figure modified from Ii et al. (2021) including the new data of 2020 and 2021]

B: Straw

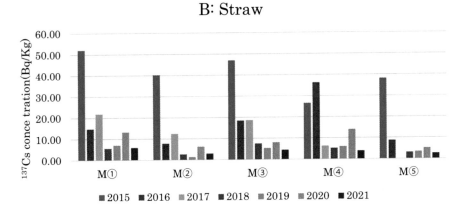

Fig. 3.5 The yearly change of ^{137}Cs in straw (**b**) at the sampling points (2015–2021). The values are adjusted to the date of measurement of the soil within a week after the rice sampling. Data are not presented for M⑤ in 2017, because the spectrum peak shape did not show Gaussian distribution. [Figure modified from Ii et al. (2021) including the new data of 2020 and 2021]

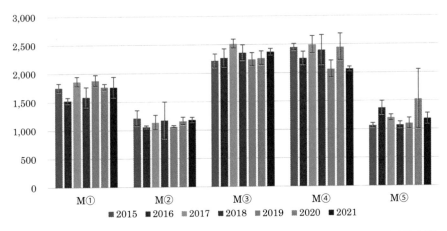

Fig. 3.6 ^{137}Cs concentration (Bq/kg of dry soil) of paddy soil at each sampling point (M①, M②, M③, M④ and M⑤) from 2015 to 2021. [Figure modified from Ii et al. (2021) including the new data of 2020 and 2021]

Fig. 3.1. Those of M③ and M④ are about 2200 Bq/kg, and that of M① is about 1700 Bq/kg, keeping the gradient of ^{137}Cs concentration at the decontamination time in 2012[1),2)]. The yearly change of the ^{137}Cs concentration from 2015 to 2021 is not obvious at each point. This means relatively slow soil ^{137}Cs movement from 2015 to 2021. Theoretically ^{137}Cs (half-life: 30.08 year) should reduce 13% during 6 years. However, the 13% reduction was not obvious because of experimental variance and due to probable introduction of some quantity of ^{137}Cs from outer contaminated fields with water and dust.

3.3.2 Yearly Change of Transfer Factors of ^{137}Cs to Brown Rice and Straw from Soil

Figures 3.7 and 3.8 show the yearly change of TF of ^{137}Cs from soil to brown rice (a) and to straw (b), respectively, which can be calculated from Figs. 3.4, 3.5 and 3.6. These show more than 80% decreases in the TF, those are brown rice: 0.0022 ± 0.0008 in 2015 to 0.0003 ± 0.0001 in 2019 and straw: 0.0262 ± 0.0102 in 2015 to 0.0028 ± 0.0009 in 2019. Little decrease was observed from 2019 to 2021.

Figure 3.9 shows semilog plots of all the data of TF of ^{137}Cs to year axis (c). Regression lines and formulas for brown rice and straw are included. Semilog plots show better fitting than usual plots (a and b). Regression analyses of the plots show that R^2 values are 0.72 for brown rice ($N = 31$) and year, and 0.60 for straw ($N = 34$) and year, which still means strong correlations between the TF and year (ageing), and the correlations are significant (P value: 1.58E−09 for brown rice and 9.15E−08 for straw). From the regression formulas, annual decreasing rates of TF are calculated to be 28% (95% confidence range: 22–33%) for brown rice and 28% (95% confidence range: 21–35%) for straw, respectively. These values are lower than the values of 39% for brown rice and 43% for straw calculated from the data of 2015 to 2019, but close to 30–40% decrease was reported by Fujimura and Eguchi (2016) employing pot experiments from 2011 to 2012, though cultivation years are much earlier. Our early results (Ii et al. 2015; Ii and Tanoi 2016) in 2012 and 2013 at the same Sasu field also showed about 30% decrease in the TF of brown rice, compared in the KCl-fertilized fields. The data of 2012–2014 were not included in this report, since the fields were divided into fields with KCl and without KCl and the sampling points were different from those of 2015 to 2021.

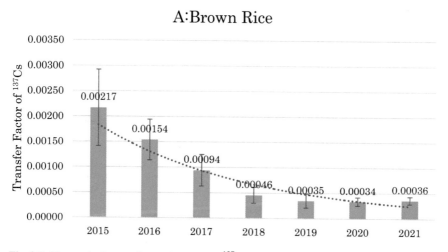

Fig. 3.7 The yearly decrease in transfer factor of ^{137}Cs from soil to brown rice (**a**) [Figure modified from Ii et al. (2021) including the new data of 2020 and 2021]

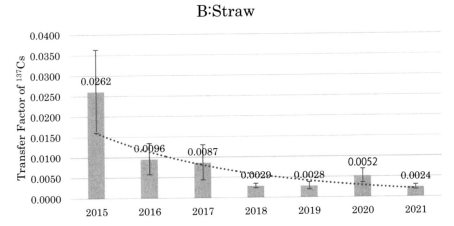

Fig. 3.8 The yearly decrease in transfer factor of ^{137}Cs from soil to straw (**b**) Figure modified from Ii et al. (2021) including new data of 2020 and 2021. Straw values (red number) in 2020 are high, probably due to sampling of more leaf part. [Figure modified from Ii et al. (2021) including the new data of 2020 and 2021]

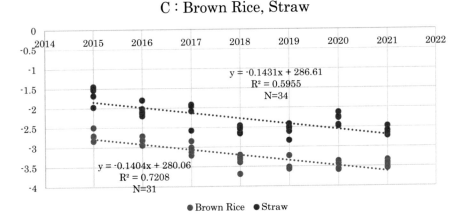

Fig. 3.9 Semilog plots of all data obtained of transfer factors of ^{137}Cs from soil to brown rice (blue dots: below) and to straw (red dots: upper) to year axis (**c**). Regression lines and formulas with N (number of data analyzed). [Figure modified from Ii et al. (2021) including new data of 2020 and 2021]

Figure 3.10 shows results of measurement of exchangeable positive ions (Na, K, Mg, and Ca) of the sampled soil. In 2015, all the ion concentrations were low, compared to 2014 and 2016–2019, this may be due to outflow of surface fine soils and sedimentation of fines flowed in with the floodwater by typhoon 18 on September 10 before the soil sampling. The exchangeable K ion increased from 14.2 ± 1.8 mg/100 g of soil (2015) to 22.8 ± 2.0 mg/100 g of soil (2019) as K_2O, and

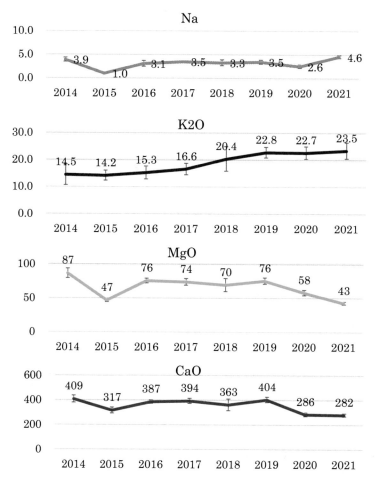

Fig. 3.10 The exchangeable cation concentrations of the soil (2014–2021). Each point is the average with a range of SD. $N = 5$ except 3 in 2015. The values are expressed as mg of Na, K_2O, MgO and CaO per 100 g of dry soil (vertical axis). [Figure modified from Ii et al. (2021) including new data of 2020 and 2021]

little change from 2019 to 2021. Since the increase of exchangeable K ion reduces radiocesium absorption by plant from soil, the increase could explain the decrease of the TF from 2015 to 2019, and the above single regression analyses of TF with year as a single variable may overestimate the decreasing rates by year. Further possibly strong relationship between year and K_2O may cause multicollinearity and multiple regression analyses may be hampered, then Variance Inflation Factor (VIF) between year and K_2O of the data sets from 2015 to 2021 was calculated to be 3.76 for brown rice ($N = 29$) and 2.54 for straw ($N = 32$), respectively, and the multicollinearity is not serious for the analyses. Multiple regression analyses of all sets of data of logarithm of TF in 2015–2021 to year and K_2O as variables were performed and

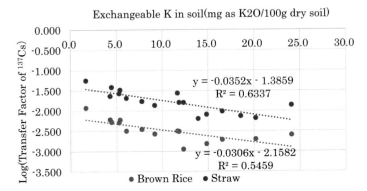

Fig. 3.11 Relation between TF and exchangeable K_2O concentration obtained at Sasu and Komiya paddy fields in Iitate Village in 2016. Straw (red points: above). Brown rice (blue points: below)

show that R^2 values are 0.73 for brown rice ($N = 29$) and 0.61 for straw ($N = 32$) with year and K_2O, and significant P values for year of 0.0002 for brown rice and of 0.0014 for straw, and significant P values for K_2O of 0.016 for brown rice but nonsignificant P value of 0.079 for straw, respectively, assuming significant P level of 0.05. From the following multiregression formulas:

Log (TF of brown rice) $= 161.704 - 0.08147$ Year $- 0.02694$ K_2O (mg/100 g of soil),

Log (TF of straw) $= 183.652 - 0.09186$ Year $- 0.02856$ K_2O (mg/100 g of soil),

annual decreasing rates of TF are calculated to be 17% for brown rice (95% confidence range, 7–26%) and 19% for straw (95% confidence range: 4–32%). These mean that the decreases in TF from 2015 to 2021 are mainly due to year (ageing) effect, but that the K_2O annual increase (1.55 mg as average) may add the decrease by 9% for brown rice and 10% for straw as a supplemental manner, the values of decrease are comparable with 8% decrease for brown rice and 9% for straw, calculated by the K_2O increase with the formulas of the relation between TF and exchangeable K_2O concentration obtained at Sasu and Komiya in 2016, as shown in Fig. 3.11.

These suggest that ^{137}Cs in soil was gradually transformed to a form more difficult to be absorbed by rice, and we assessed possible annual change of exchangeable ^{137}Cs fraction in the soil.

3.3.3 Analysis of Exchangeable ^{137}Cs Fraction in the Soil Sampled in 2019, 2020 and 2021

In a previous report (Ii et al. 2021), we compared the fraction of exchangeable ^{137}Cs to whole ^{137}Cs in the soil sampled in 2017, 2018, and 2019 at each sampling point. At all the points, the exchangeable ^{137}Cs fraction decreased from 2017 to 2019 and

A

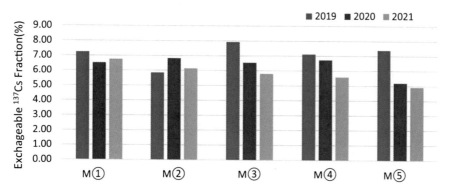

Fig. 3.12 (a) Comparison of the fraction (%) of exchangeable ^{137}Cs in the soil at the sampling points (2019–2021)

multiple regression analysis of the set of the data with year and K_2O showed that annual decrease rate of exchangeable fraction is 13% (95% confidence range: 0.4–24%) and annual decrease fraction by the annual increase of K_2O (3.1 mg/ 100 g of soil) could be 4%. A similar analysis was performed using the soil samples of 2019, 2020, and 2021.

Figure 3.12a shows of exchangeable ^{137}Cs fraction (%) in the soil at the sampling points from 2019 to 2021. Above 90% ^{137}Cs of the paddy soil is not exchangeable, that is, not extracted with 1 M CH_3COONH_4 solution. Figure 3.13 shows the regression analysis of all the data to year axis, a little decrease of co. 0.6% per year observed in exchangeable ^{137}Cs fraction. R^2 value of 0.40 with P value of 0.011. However, almost no change in the TF for brown rice and straw from 2019 to 2021.

The exchangeable K_2O changed little from 2019 to 2021 (23 mg/100 g of dry soil) as shown in Fig. 3.10, and multiple regression analysis was not performed this time. The decrease appeared to reach close to the bottom level.

The annual decrease rate by year of 13% from 2015 to 2019 is much less than those of the TF of about 40%, suggesting other factor(s) not identified concerned. Further, this is less than the annual decrease rate of 34% calculated from data in Tsukada's report (2014) of a paddy field in Date-city in Fukushima during 1.5 years from 2012 to 2013, but more than that of 4–10% calculated from the half-time of 6.6–17.7 year of ^{137}Cs/^{133}Cs in the exchangeable fraction, using the data from 2011 to 2015, as better index of ^{137}Cs fixation by ageing (Wakabayashi et al. 2020).

This consecutive field work from 2015 to 2021 indicates that ^{137}Cs in the soil was gradually transformed to a form more difficult to be absorbed by rice, that is considered due to gradual radiocesium fixation with ageing, as shown by Roig et al. (2007), Absalom et al. (1995), Yamaguchi et al. (2016, 2019) and Takeda et al. (2013, 2020) who investigated fixation mechanism and conditions in detail

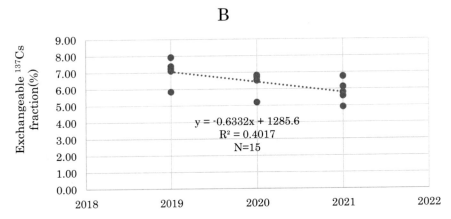

Fig. 3.13 (**b**) Plot and analysis of all the data of exchangeable ^{137}Cs fraction (%) in the soil to year axis in (**a**)

mainly under artificial conditions. However, their data are from those within 3 years from radiocesium fall-out or addition to soils. Our work showed that phytoavailability of ^{137}Cs assessed by TF of rice and exchangeable ^{137}Cs fraction of the soil continue to decrease even after 4–8 years from the fall-out in an actual paddy field in Iitate Village in Fukushima Prefecture This slow decrease may be partly due to the type of soil of the allophanic Andosol of the test field, lower ageing speed, compared to mineral soil (Absalom et al. 1995; Yamaguchi et al. 2019), and also due to probable presence of ^{137}Cs in organic materials not exchangeable in soil (co. 6% in case of Date-city soil, Tsukada 2014) and ^{137}Cs in harvest residue returned and incorporated outer organic matters, which may gradually be transformed exchangeable and further fixed to mineral content in the soil. The mechanism remains to be clarified.

3.4   Conclusive Remark

Just after the disaster of FDNPP, we decontaminated the paddy field by removing the contaminated surface soil by mixing the surface (0–5 cm) and flow-out. During the period between 2012 and 2014, we reconfirmed the importance of KCl addition to reduce radiocesium concentration of rice, and could clear the safety standard for food in Japan (<100 Bq/kg). Natural reduction of radioactivity with decay of ^{134}Cs (half-life of 2 years) helped the reduction by 75% for 4 years by 2015. However, we cannot further expect such natural reduction of radioactivity with decay of ^{137}Cs (half-life of 30 years). Nevertheless, consecutive yearly reduction of radiocesium activity in harvested rice was observed (2015–2019) at a KCl-fertilized and straw plowed-in paddy field and this is suggested mainly due to gradual radiocesium fixation by soil with ageing and further reduction of radiocesium contamination in rice was expected. However, the following study of 2019–2021 suggests that the TF of brown rice and straw may almost reach the bottom values of 0.00035 for brown rice and 0.002 for straw (Figs. 3.7 and 3.8). Further this field work also shows the robustness of the rice cultivation that we could annually harvest rice far below the safety standard, even though the paddy field suffered floods by typhoons in September 2015 and in October 2019.

We wish to extend rice cultivation to other paddy fields and sooner resurrection of agriculture in Fukushima.

Acknowledgements I acknowledge Mr. Muneo Kanno in Sasu and Mr. Kin-ichi Ookubo in Komiya in the Iitate Village for cooperation of the rice cultivation, members of NPO Resurrection of Fukushima (representative: Mr. Yoichi Tao), and of "Circle Madei" (representative: Ms. Tomiko Saito) at The University of Tokyo, and Dr. Naoto Nihei of Department of Agriculture, Fukushima University, Dr. Atsushi Hirose of Hoshi University, Dr. Natsuko Kobayashi and Dr. Masaru Mizoguchi for the collaboration of this research and Ms. M. Katsuno and Ms. M. Takemura in Graduate school of Agricultural and Life Sciences, The University of Tokyo for the assistance of radioactivity measurement and chemical analyses.

This study is partially supported by the Mitsui & Co. Environment Fund K12-F-903 (2012–2015) to NPO Resurrection of Fukushima and a JMAFF grant in 2016 to The University of Tokyo (Rep. Dr. Naoto Nihei).

References

Absalom JP, Young SD, Crout NMJ (1995) Radiocaesium fixation dynamics: measurement in six Cumbrian soils. Eur J Soil Sci 46:461–469

Fujimura S, Eguchi T (2016) Transfer factors of radiocesium to brown rice in the decontaminated paddy field. Jpn J Crop Sci 85:211–217

Ii I, Tanoi K (2016) Consecutive field trials of rice cultivation in partially decontaminated paddy fields to reduce radiocesium absorption in the Iitate village in Fukushima prefecture. In: Nakanishi TM, Tanoi K (eds) Agricultural implications of the Fukushima nuclear accident. Springer, Tokyo, pp 55–76

Ii I, Tanoi KT, Uno Y, Nobori T, Hirose A, Kobayashi N, Nihei N, Ogawa T, Tao Y, Kanno M, Nishiwaki J, Mizoguchi M (2015) Radioactive cesium concentration of lowland rice grown in the decontaminated fields in Iitate-Village in Fukushima. Radioisotopes 64:299–310

Ii I, Nihei N, Hirose A, Kobayashi N et al (2021) Annual reduction of transfer factor of radiocesium from soil to rice cultivated in a KCl fertilized paddy field from 2015 to 2019. Radioisotopes 70: 63–72

Nobori T, Tanoi K, Nakanishi TM (2013) Method of radiocesium determination in soil and crops using a NaI(Tl) scintillation counter attached with an autosampler. Jpn J Soil Sci Plant Nutr 84: 182–186

Ono Y, Sato K, Sakuma H, Nemoto K et al (2013) Variation in rice radiocesium absorption among different cultivars. Research Report of Fukushima Agricultural Technology Center: Special Issue for Counteraction to Radioactive Materials, 29–32

Roig M, Vidal M, Rauret G, Rigol A (2007) Prediction of radionuclide aging in soils from the Chernobyl and Mediterranean areas. J Environ Qual 36:943–952

Tagami K, Tsukada H, Uchida S, Howard BJ (2018) Changes in the soil to brown rice concentration ratio of radiocaesium before and after the Fukushima Daiichi Nuclear Power Plant accident in 2011. Environ Sci Technol 52:8339–8345

Takeda A, Tsukada H, Nakao A, Takaku Y et al (2013) Time-dependent changes of phytoavailability of Cs added to allophanic andosols in laboratory cultivations and extraction. J Environ Radioact 122:29–36

Takeda A, Tsukada H, Unno Y, Takaku Y et al (2020) Effects of organic amendments on the natural attenuation of radiocesium transferability in grassland soils with high potassium fertility. J Environ Radioact 217:106207

Tsukada H (2014) Behavior of radioactive cesium in soil with aging. Jpn J Soil Sci Plant Nutr 85: 77–79

Wakabayashi S, Takahashi S, Matsunami H, Hamamatsu S et al (2020) Evaluation of ^{137}Cs/^{133}Cs ratio in andosol paddy fields with/without potassium fertilizer application. J Environ Radioact 218:106252

Yamaguchi N, Taniyama I, Kimura T, Yoshioka K et al (2016) Contamination of agricultural products and soils with radiocesium derived from the accident at TEPCO Fukushima Daiichi Nuclear Power Station: monitoring, case studies and countermeasures. Soil Sci Plant Nutr 62: 303–314

Yamaguchi N, Hikono A, Saito T (2019) Effects of zeolite and vermiculite addition on exchangeable radiocaesium in soil with accelerated ageing. J Environ Radioact 203:18–24

Yamamura K, Fujimura S, Ota T, Ishikawa T et al (2018) A statistical model for estimating the radiocesium transfer factor from soil to brown rice using the soil exchangeable potassium content. J Environ Radioact 195:114–125

Yamashita M, Eguchi S, Tateishi T, Tsuiki M (2016) Species difference and yearly change of radioactive cesium concentration in grasses. Jpn J Grassl Sci 62:134–139

Chapter 4
Effects of Radioactive Cesium from Suspended Matter and Fallout on Agricultural Products

Naoto Nihei and Kazuya Yoshimura

4.1 Introduction

The disastrous accident at the Tokyo Electric Power Company's Fukushima Daiichi Nuclear Power Plant (FDNPP) in 2011 caused the release of significant amounts of radioactive waste, resulting in radiological contamination in the surrounding forests, agricultural fields, and residential areas. Although ^{131}I and ^{132}I were the main radionuclides found immediately after the accident, the environmental radiation was primarily derived from radioactive cesium (Cs), particularly ^{134}Cs and ^{137}Cs from 10 days after the accident (United Nations Scientific Committee on the Effects of Atomic Radiation 2014; Yoshimura et al. 2020). Radioactive Cs was released either in gaseous or insoluble particulate form, and the gaseous form was considered to be transported with sulfate-suspended matters in the atmosphere, deposited on the ground, and attached to soil particles or vegetation (Kaneyasu et al. 2012). Radioactive Cs activity in the atmosphere has gradually decreased since the FDNPP accident (Abe et al. 2021). However, it can still be detected in fallout and atmospheric suspended matter, especially in areas around the FDNPP, including the decontaminated fields (Nuclear Regulation Authority, Japan 2019a, b). One of the contributing factors could be a resuspension of radioactive Cs deposited on the ground (Ishizuka et al. 2017).

Agriculture is the main industry in the Fukushima Prefecture. To produce safe agricultural products, field decontamination, potassium fertilization to prevent

N. Nihei (✉)
Faculty of Food and Agricultural Sciences, Fukushima University, Fukushima City, Fukushima, Japan
e-mail: nihei@agri.fuhukushima-u.ac.jp

K. Yoshimura
Sector of Fukushima Research and Development, Japan Atomic Energy Agency, Minamisoma City, Fukushima, Japan

© The Author(s) 2023
T. M. Nakanishi, K. Tanoi (eds.), *Agricultural Implications of Fukushima Nuclear Accident (IV)*, https://doi.org/10.1007/978-981-19-9361-9_4

absorption of radioactive Cs via roots, and other counter measures have been considered (Fukushima Prefecture 2014; Ministry of the Environment 2013). However, secondary contamination of agricultural products by radioactive cesium via fallout and suspended matter has not been contemplated in these counter measures. Therefore, it is important to understand the effect of radioactive cesium from fallout and suspended matter on the secondary contamination of agricultural products.

4.2 Effects of Radioactive Cs from Suspended Matter and Fallout on Japanese Mustard Spinach (Komatsuna)

To determine the effects of radioactive Cs from suspended matter and fallout on the secondary contamination of agricultural products, Japanese mustard spinach (*Brassica rapa* L. var. *perviridis,* Komatsuna) was cultivated in the area affected by the FDNPP accident (Nihei et al. 2018). Briefly, Komatsuna was cultivated using noncontaminated soil and water. Pots containing Komatsuna were placed 30 cm (hereafter referred to as "below") or 120 cm (referred to as "above") above the ground surface (Fig. 4.1). The pots were placed at two locations in the Fukushima Prefecture: Site A, approximately 35 km from the FDNPP with an air dose rate of approximately 0.4 μSv h^{-1} (measured in August 2018), and Site B, approximately 3.5 km from the FDNPP with an air dose rate of 10.0 μSv h^{-1}. Site A was in a zone that was evacuated until March 2016 and subsequently decontaminated. Site B was

Fig. 4.1 Komatsuna cultivation system installed at the test site. Pots filled with noncontaminated soil were placed at heights of 30 cm and 120 cm above the ground. Basins filled with water were placed to collect fallout at the same heights as the pots

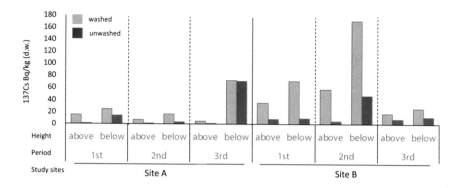

Fig. 4.2 Radioactive Cs concentration of Komatsuna cultivated using noncontaminated soil in each site

also within an evacuation zone, but was not yet decontaminated. Komatsuna were cultivated during three periods: August 16–September 9, 2018 (Period 1), September 9–October 6, 2018 (Period 2), and October 6–November 23, 2018 (Period 3). After harvesting part of the Komatsuna, half of the plants were not washed (hereafter, "unwashed plants") and the other half were washed with water ("washed plants"). To collect the fallout, basins filled with water were placed at each site during each cultivation period. Radiocesium in the fallout was collected as dissolved and particulate fractions, which were separated by filtration. Activity of ^{137}Cs in the sample was measured using a Ge-semiconductor detector.

Figure 4.2 shows the ^{137}Cs concentrations in the unwashed and washed Komatsuna samples cultivated at each study site. The measurement results are expressed as activity concentration of ^{137}Cs per kg of dry weight (Bq kg^{-1}). The dry weight-base concentration is approximately ten times higher than the wet weight-base concentration, which is applied to official monitoring inspections of agricultural products. The concentration of ^{137}Cs in Komatsuna cultivated at Site A ranged from 8 to 73 Bq kg^{-1}, while that at Site B showed higher values ranging from 18 to 171 Bq kg^{-1}, with secondary contamination depending on the ^{137}Cs level in the fields. If we consider the wet weight-base concentration, the values were less than the standard limit of radioactive Cs concentration allowed for agricultural products (100 Bq kg^{-1}) for shipping and ingestion.

The ^{137}Cs concentrations of Komatsuna cultivated at different heights showed that closer proximity to the ground generally related to higher concentrations of ^{137}Cs (Fig. 4.2). Similarly, Fig. 4.3 shows increased ^{137}Cs concentrations in the fallout collected in lower basins than that in higher basins. This difference in the ^{137}Cs concentration depending on installation height suggests that deposition of soil particles that rebounded from the ground due to rainfall contributed to secondary contamination of the Komatsuna.

Although the ^{137}Cs concentration in washed plants was significantly lower than that in unwashed plants, the washed plants showed detectable activity. This suggests that the secondary contamination was caused not only by external adhesion to the

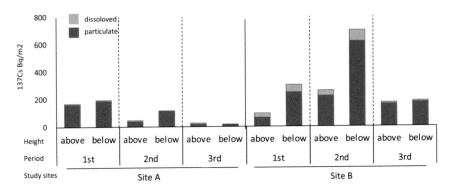

Fig. 4.3 Deposition of ^{137}Cs in the water samples from the basins installed at Sites A and B

leaf surface but also internal absorption of ^{137}Cs from the suspended matter or fallout. Assuming that the ^{137}Cs concentration in the unwashed plants was a result of both external adhesion on the leaf surface and internal absorption, and that the ^{137}Cs concentration in washed plants was derived from internal absorption only, the proportion of internal absorption to total ^{137}Cs was estimated as the ratio between the ^{137}Cs concentrations of washed and unwashed plants. The average proportions were estimated to be 62% at Site A and 71% at Site B, with variation depending on the cultivation period.

4.3 Secondary Contamination Due to Fallout

There are two possible pathways for secondary contamination of agricultural products due to fallout: root uptake of radioactive Cs deposited on the soil in the pots, or foliar absorption of radioactive Cs adhered to the plant surface. To investigate the root uptake, the pots were filled with noncontaminated soil and left for 31, 77, and 161 days in Site B. After each period, the soil in the pot was used to cultivate Komatsuna in a phytotron. The ^{137}Cs concentrations of the soil at 31, 77, and 161 days were 3700, 4500, and 7000 Bq m^{-2}, respectively, and the ^{137}Cs concentrations of the Komatsuna grown in the respective soils were 9, 21, and 34 Bq kg^{-1}. These results indicate that root uptake of radioactive Cs from the contaminated soil due to fallout is a legitimate pathway for secondary contamination of agricultural products. However, the fact that the radioactive Cs concentration in Komatsuna was below the allowable limit for shipping and ingestion, despite the ^{137}Cs concentration in soil reaching 4500 Bq kg^{-1} by staying in the evacuation zone, suggests that the impact of this pathway on secondary contamination is minimal.

4.4 Characteristics of Radioactive Fallout

The ^{137}Cs fallout at each site showed that approximately 90% of the ^{137}Cs was detected in the particulate fraction, while the remaining 10% was in soluble form (Fig. 4.3). Part of the dissolved Cs could present in a particulate form during suspension in the atmosphere and then dissolved in the water of basins during the fallout-collection periods. This dissolved fraction is bioavailable and could be absorbed from the attached leaf surface or the roots.

To analyze the fallout particulate matter in detail, the particulates in the basin at Site B were collected on membrane filters and analyzed using autoradiography. The autoradiographic images of the membrane filter showed black dots, which represent the presence of radioactive particles (Okumura et al. 2019) in these samples (Fig. 4.4). One of the particles had a diameter of approximately 2 μm and radiation of approximately 0.5 Bq. Particles were investigated using a scanning electron microscope with an energy-dispersive X-ray spectrometer (EDS). The peaks of Si and O suggest that the matter consists mainly of silicate. Additionally, Fe, Zn, K, Cl, and Cs were detected. These characteristics of radioactive particles are similar to those of the Cs-bearing microparticles reported by Adachi et al. (2013). These particles containing high concentrations of radioactive Cs rarely increase the radioactive Cs in agricultural products, because it is difficult to leach the cesium from them. The other particles had approximate diameters of 60–120 μm and approximate radiation levels of 0.4–1.2 Bq. EDS spectra showed an abundance of Si, Al, and O, suggesting that these are basically soil particles with fine clay minerals. Since radioactive Cs that is stably fixed in the clay layer is difficult to dissolve, the radioactive Cs in these particles is likely only minimally absorbed by plants.

Fig. 4.4 Autoradiograph image of radioactive materials collected from water samples and detected with imaging plate. Black dots indicate presence of radioactive materials

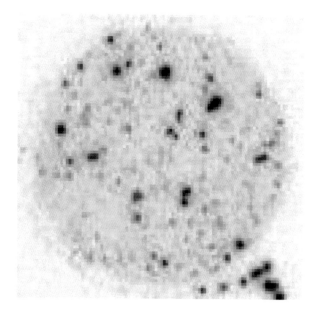

However, in addition to the glassy or clay-fixed form observed in this study, radioactive Cs in suspension also adsorb onto organic matter (Okumura et al. 2019). Organic matter could decompose over time and release radioactive Cs. Therefore, further investigation is necessary regarding the bioavailability of radioactive Cs in organic particles and their effect on secondary contamination of plants.

4.5 Relationship Between Fallout Adhesion and Meteorological Factors

The concentration of ^{137}Cs in Komatsuna fluctuated depending on the cultivation period within the same area. Factors such as wind speed and precipitation were examined to assess their effect on the varying ^{137}Cs concentrations observed during the growth periods of Komatsuna (Fig. 4.5). No clear relationship between these factors and ^{137}Cs concentration was observed for Site A, but higher precipitation and faster maximum wind speeds resulted in higher ^{137}Cs concentrations in Site B for Komatsuna. Particles containing radioactive Cs deposited on soil and vegetation are considered to have been carried by wind and precipitated by rain (Nuclear Regulation Authority, Japan 2019a). These meteorological factors, in addition to the types and growth stage of plants, multiply the effects on the adsorption and detachment of radioactive Cs during cultivation, likely causing the inconsistent results between Sites A and B. Therefore, further research is needed to quantitatively evaluate the effects of these factors on the activity of radioactive Cs in agricultural products.

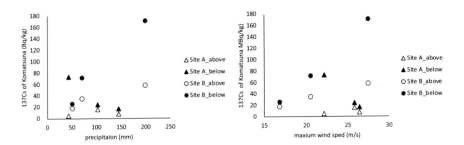

Fig. 4.5 Correlation analysis between precipitation, maximum wind speed, and radioactive Cs levels in cultivated Komatsuna

4.6   Conclusions

To determine the effects of radioactive Cs in suspended matter and fallout on the secondary contamination of agricultural products, Komatsuna was cultivated using noncontaminated soil and water at sites affected by the FDNPP accident. Komatsuna cultivated at the decontaminated site had lower concentrations of radioactive Cs than those cultivated at the radioactive-contaminated site. Komatsuna cultivated closer to the ground showed higher levels of radioactive Cs than those cultivated at greater heights. This result suggests that the radioactive Cs of Komatsuna grown in noncontaminated soil is affected by suspended matter as well as the deposition of soil particles that rebounded from the ground due to rainfall. The levels of radioactive Cs decreased when the plants were washed with water. This result suggests that the secondary contamination of Komatsuna was derived from both absorption into the plant and adhesion to the plant surfaces. The fallout contained both dissolved and particulate forms of radioactive Cs, and it was assumed that a portion of these contributed to secondary contamination. To reduce the risk in the future, secondary contamination can be mitigated through the understanding of environmental and meteorological factors that increase the contamination risk, secondary contamination sources, and contamination differences by crop type.

References

Abe T, Yoshimura K, Sanada Y (2021) Temporal change in atmospheric radiocesium during the first 7 years after the Fukushima Dai-ich nuclear power plant accident. Aerosol Air Qual Res 21(7):200636. https://doi.org/10.4209/aaqr.200636

Adachi K, Kajino M, Zairzen Y, Igarashi Y (2013) Emission of spherical cesium-bearing particles from an early stage of the Fukushima nuclear accident. Sci Rep 3:2554. https://doi.org/10.1038/srep02554

Fukushima Prefecture (2014). https://www.pref.fukushima.lg.jp/uploaded/attachment/61508.pdf

Ishizuka M et al (2017) Use of a size-resolved 1-D resuspension scheme to evaluate suspended radioactive material associated with mineral dust particles from the ground surface. J Environ Radioact 166:436–448

Kaneyasu N, Ohashi H, Suzuki F, Okuda T, Ikemori F (2012) Sulfate suspended matter as a potential transport medium of radiocesium from the Fukushima nuclear accident. Environ Sci Technol 46:5720–5726

Ministry of the Environment (2013) Decontamination guidelines, 2nd ed. http://josen.env.go.jp/en/framework/pdf/decontamination_guidelines_2nd.pdf

Nihei N, Yoshimura K, Okumura T, Tanoi K, Iijima K, Kogure T, Nakanishi TM (2018) Secondary radiocesium contamination of agricultural products by suspended matter. J Radioanal Nucl Chem 318:341–346

Nuclear Regulation Authority, Japan (2019a) Monitoring information of webpage (fallout). https://radioactivity.nsr.go.jp/en/list/280/list-1.html

Nuclear Regulation Authority, Japan (2019b) Monitoring information of webpage (suspended matter). https://radioactivity.nsr.go.jp/en/list/200/list-1.html

Okumura T et al (2019) Dissolution behavior of radiocaesium-bearing microparticles released from the Fukushima nuclear plant. Sci Rep 9(1):3520

United Nations Scientific Committee on the Effects of Atomic Radiation (2014) Annex, A. Levels and effects of radiation exposure due to the nuclear accident after the 2011 great east-Japan earthquake and tsunami, 2014. UNSCEAR 2013 Report. Sources, effects and risks of ionizing radiation. UNSCEAR, New York

Yoshimura K, Saegusa J, Sanada Y (2020) Initial decrease in the ambient dose equivalent rate after the Fukushima accident and its difference from Chernobyl. Sci Rep 10:3859

Chapter 5
Verification of Uptake and Transport Properties of Cesium in Hydroponically Cultivated *Quercus serrata*

Natsuko I. Kobayashi and Riona Kobayashi

5.1 Introduction

Under natural environment where various incidents such as drought, flooding, mycorrhizal symbiosis, and pathogen invasion can occur, plants acquire minerals and thrive. Such a complex and fluctuating growing environment cannot be reproduced by an artificial hydroponic cultivation system, which is generally employed in plant science laboratories. However, it is also true that experiment using hydroponic medium containing essential minerals contributes to the investigation of some fundamental physiological characteristics of plants. The mechanism by which plant roots selectively uptake some ions from soils containing various concentrations of many elements has been investigated actively in 1900s. For potassium uptake, detailed kinetic analysis using liquid medium suggested that it is the carrier site present in the roots that allow for potassium ion (K^+)-selective uptake (Epstein et al. 1963). For the K^+ transporters that were subsequently identified, molecular biological studies are now progressing to understand the mechanisms that confer K^+ selectivity on them.

Since the recognition of radiocesium as an environmental contaminant, the relationship between K^+ and Cs^+ absorption has been investigated using hydroponics, with the goal of reducing the absorption of radiocesium by plants (Zhu and Smolders 2000). The results of many previous studies, including the inverse proportional relationship between K^+ supply and Cs^+ uptake, and the identification of the Cs^+-permeable K^+ transporter *Arabidopsis* HAK5 (Qi et al. 2008), have provided an important scientific basis for the effect of potassium fertilization in crop production on contaminated agricultural lands after the Fukushima accident. Indeed,

N. I. Kobayashi (✉) · R. Kobayashi
Graduate School of Agricultural and Life Sciences, The University of Tokyo, Tokyo, Japan
e-mail: anikoba@g.ecc.u-tokyo.ac.jp

© The Author(s) 2023
T. M. Nakanishi, K. Tanoi (eds.), *Agricultural Implications of Fukushima Nuclear Accident (IV)*, https://doi.org/10.1007/978-981-19-9361-9_5

potassium fertilization has been very successful in reducing radiocesium contamination in agricultural products (Kato et al. 2015; Kubo et al. 2017).

However, studies that have contributed to accumulating basic knowledge through hydroponic systems have mainly been conducted using herbaceous plants that are easy to handle in the laboratory. Do the trees contaminated with radiocesium in the forests of Fukushima have the same mechanisms of uptake and transport of cesium that we know about? One way to answer this question would be to apply the classic experiment of growing plants in an environment with controlled K^+ and Cs^+ concentrations to investigate the content of both elements in the plant body to trees. Therefore, we conducted a study using hydroponically grown Konara oak (*Quercus serrata* Thunb.) to analyze the characteristics of Cs^+ uptake in the context of its relationship with K^+ (Kobayashi et al. 2019).

5.2   The Effect of K Nutrition on the Cs Content

Oak seeds harvested from a single Konara oak tree were incubated in vermiculite for germination and further cultivated in the hydroponic medium for raising seedlings. Then, the 4-seek-old oak seedlings were transferred to the hydroponic medium containing 50, 200, or 3000 μM K^+ and 0.1 μM Cs^+ (Table 5.1, Kobayashi et al. 2019). After 4 weeks, the content of K and Cs in the seedling developing 4 leaves at the top (upper leaves) and another 4 at the bottom node (lower leaves) was analyzed using the inductively coupled plasma-mass spectrometry, and it was found that Cs content in the seedlings cultivated with 3000 μM K^+ was lower than other seedlings (Table 5.1, Kobayashi et al. 2019). The difference, however, was not as large as we had expected. The Cs content in the shoot tissues of the seedlings grown with 50 μM K^+ was only twice as high as that in the seedlings grown with 3000 μM K^+ (Table 5.1, Kobayashi et al. 2019). This is markedly different from what we know from similar experiments conducted on several herbaceous plants, which showed that plant Cs contents can drop below one-tenth when the K concentration in the nutrient solution increases from 50 to 1000 μM (Zhu and Smolders 2000). The K content in oak seedlings was increased as the K^+ concentration in the hydroponic medium increased (Table 5.1, Kobayashi et al. 2019). The difference in the shoot K content between cultivating under 3000 μM K^+ condition and under 50 μM K^+ condition was almost twofold (Table 5.1, Kobayashi et al. 2019). Similar observation was reported for the soybean plants grown under 30 and 3000 μM K^+; fourfold difference in shoots and twofold difference in roots (Nihei et al. 2018). Plants in general will have a mechanism to keep K content stable under conditions in which the K^+ concentration in the culture medium fluctuates in the 100-fold range.

To understand the characteristics of Cs^+ dynamics in plants, the discrimination factor (DF) based on the ratio between K and Cs has been frequently referred. Here, we calculated K/Cs discrimination factor for oak seedlings by dividing K/Cs value in plant tissue by K/Cs value in the culture medium (Table 5.2). The result was what could be called unexpected. Under 200 μM K^+ condition, the K/Cs DF was almost

Table 5.1 Growth and elemental content of 8-week-old oak seedlings grown in the preculture medium containing K^+ at various concentrations and 0.1 μM Cs^+ ($n = 6$–11) (Kobayashi et al. 2019)

Cultivation		Weight (g) Whole plant	Cs content (μg/g)					K content (mg/g)				
			Upper leaves	Lower leaves	Stems	Coarse roots	Fine roots	Upper leaves	Lower leaves	Stems	Coarse roots	Fine roots
50 μM K	Average	0.82	3.08	2.15	2.12	4.15	35.37	4.72	3.40	4.38	4.78	10.87
	S.D.	0.17	0.67	0.64	0.35	1.23	18.97	1.12	0.63	0.68	0.70	2.86
200 μM K	Average	0.68	2.66	1.79	1.94	3.17	8.59	5.33	3.92	5.31	6.16	13.76
	S.D.	0.25	0.36	0.36	0.49	0.32	1.73	1.34	0.64	0.73	0.66	1.68
3000 μM K	Average	0.72	1.70	0.93	1.12	1.49	2.90	9.65	7.11	8.26	8.36	15.60
	S.D.	0.20	0.22	0.20	0.15	0.17	0.41	1.08	1.25	1.19	0.81	4.78

Table 5.2 K/Cs discrimination factor (DF) values in hydroponically grown oak seedlings

Cultivation		K/Cs Discrimination factor				
		Upper leaves	Lower leaves	Stems	Coarse roots	Fine roots
50 µM K	Average	**3.13**	**3.33**	**4.19**	**2.45**	**0.70**
	S.D.	0.80	0.83	0.68	0.71	0.22
200 µM K	Average	**0.99**	**1.12**	**1.41**	**0.98**	**0.85**
	S.D.	0.12	0.20	0.24	0.11	0.27
3000 µM K	Average	**0.19**	**0.26**	**0.25**	**0.19**	**0.18**
	S.D.	0.02	0.03	0.03	0.02	0.04

DF values were calculated by dividing the ratio of K and Cs content in each organ (Table 5.1) by the K/Cs concentration ratio in the preculture medium

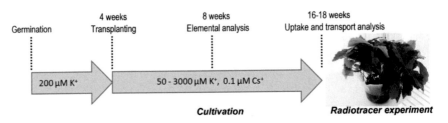

Fig. 5.1 Experimental procedures. Oak seedlings were grown hydroponically in the plant growth chamber with a controlled day/night cycle of 12 h/12 h and 25 °C/20 °C (Kobayashi et al. 2019)

1 in all tissues (Table 5.2), which means that Cs^+ and K^+ are absorbed and accumulated without discrimination. The K/Cs DF was further lowered when oak seedlings were cultivated under 3000 µM K^+ condition (Table 5.2). Oak seedlings never preferentially absorb K^+ than Cs^+ under K-rich environments. Such results have not been observed for herbaceous plants (Zhu and Smolders 2000). In addition, the result that the lower the K^+ concentration in the medium, the higher the K/Cs DF value also does not fit with previous findings. In *Arabidopsis*, K-starvation activates HAK5-mediated Cs^+ and K^+ absorption, which causes increased Cs^+ and K^+ uptake rate while decreasing K^+ selectivity in root uptake process (Qi et al. 2008). In K-rich environment, HAK5 expression is reduced and K^+ uptake comes to largely depend on the K^+-selective channel, AKT1 (Rubio et al. 2010), resulting in the elevated K/Cs DF value. To investigate whether such a shift in the K^+–Cs^+ uptake mechanism in response to K nutrition occurs in oak seedlings, we conducted an ion uptake experiment using radiotracers (Fig. 5.1). The result was that neither K^+ nor Cs^+ uptake in 2 h from the medium containing 50 µM K^+ and 0.1 µM Cs^+ differed significantly by the K nutrition in the hydroponic medium in which they were grown for more than 3 months (Table 5.3, Kobayashi et al. 2019). However, when the ratio between the K^+ uptake amount and Cs^+ uptake amount was calculated for each individual sample, it was found the value was slightly lower in the seedlings grown under 3000 µM K^+ condition (Kobayashi et al. 2019). Therefore, it seems that the

Table 5.3 The effect of K concentration in the preculture medium on the uptake and transport of Cs^+ and K^+

Cultivation		Cs$^+$ accumulation in 2 h (pmol/g)			K$^+$ accumulation in 2 h (nmol/g)		
		Shoot	Coarse roots	Fine roots	Shoot	Coarse roots	Fine roots
50 μM K	Average	1.4	106.5	229.5	1.4	46.4	120.3
	S.D.	1.0	50.1	107.9	0.5	20.1	56.7
200 μM K	Average	2.5	85.9	175.7	2.1	37.5	74.9
	S.D.	3.2	24.7	105.9	1.9	12.3	7.6
3000 μM K	Average	1.2	103.3	319.5	2.0	30.7	91.1
	S.D.	0.7	44.4	135.0	–	10.6	45.9

Oak seedlings precultured to 16–18 weeks in the medium containing 50, 200, or 3000 μM K and 0.1 μM Cs were soaked in the uptake medium containing 50 μM K, 0.1 μM Cs, and radiotracers ^{42}K and ^{137}Cs for 2 h (Kobayashi et al. 2019). The amount of Cs^+ and K^+ accumulated in various oak tissues in 2 h was calculated based on the radiation of ^{42}K and ^{137}Cs, respectively ($n = 4$–5). The signal of ^{42}K in the shoot tissue was quantitatively determined only in one sample precultured with 3000 μM K medium

K^+–Cs^+ uptake mechanism changes a little in response to K nutrition in the opposite direction of that seen in *Arabidopsis*.

5.3 Inhibition of Cs^+ Uptake Through Competition Between K^+ and Cs^+

Increased K^+ concentration in the medium can competitively reduce Cs^+ uptake (Handley and Overstreet 1961), which can be another mechanism in which the effect of K nutrition on decreasing Cs accumulation is achieved. The competition between Cs^+ and K^+ in uptake process was evaluated using oak seedlings grown under 200 μM K^+ and 0.1 μM Cs^+. We incubated these seedlings with ^{42}K and ^{137}Cs radiotracers with either 50, 200, or 3000 μM K^+ and 0.1 μM Cs^+ for 2 h, and calculated the amount of Cs^+ and K^+ uptake based on the radioactivity in each tissue (Table 5.4, Kobayashi et al. 2019). This experiment showed that 3000 μM K^+ in the medium can competitively reduce Cs^+ uptake by one-third (Table 5.4, Kobayashi et al. 2019). The intensity of competition between the two ions in oak seems to be roughly comparable to that in rice (*Oryza sativa*). Similar radiotracer experiment conducted with rice seedlings demonstrated that Cs^+ uptake rate drops by one-fifth as the K^+ concentration in the medium increased from 100 to 1000 μM (Kobayashi et al. 2016).

Table 5.4 The effect of K concentration in the uptake medium on the uptake and transport of Cs^+ and K^+

Uptake medium		Cs+ accumulation in 2 h (pmol/g)			K+ accumulation in 2 h (nmol/g)	
		Shoot	Coarse roots	Fine roots	Coarse roots	Fine roots
50 μM K	Average	2.5	91.0	273.7	49.8	186.5
	S.D.	1.8	33.9	123.2	27.0	64.1
200 μM K	Average	3.5	122.4	201.9	181.1	497.8
	S.D.	2.9	54.7	59.1	68.9	99.2
3000 μM K	Average	0.4	52.1	89.2	922.2	1040.1
	S.D.	0.2	27.6	35.9	242.3	615.0

Oak seedlings precultured for 16–18 weeks with the medium having 200 μM K and 0.1 μM Cs were soaked in the uptake medium containing either 50, 200, or 3000 μM K and 0.1 μM Cs, and radiotracers ^{42}K and ^{137}Cs for 2 h (Kobayashi et al. 2019). The amount of Cs^+ and K^+ accumulated in various Oak tissues in 2 h was calculated based on the radiation of ^{42}K and ^{137}Cs, respectively ($n = 4$–5). The signal of ^{42}K was not quantitatively determined in the shoot tissue

5.4 Uptake and Transport of Cs^+ in Oak and Rice Plants

Radiotracer experiment can describe the characteristics of ion uptake and further transport inside the plant quantitatively. Application of ^{42}K and ^{137}Cs at the same time allows simultaneous tracking of K^+ and Cs^+ and calculating the $K^+:Cs^+$ ratio for each tissue of each individual plant. This is very effective to detect any differences in the transport process of both ions. Consequently, a selective K^+ uptake and root-to-shoot transport over Cs^+ was clearly demonstrated for the rice plant grown under 274 μM K^+ and 0.1 μM Cs^+. The $K^+:Cs^+$ ratios for the rice root and leaves were approximately 5-times and 20-times higher than in the medium, respectively (Fig. 5.2). In contrast, the $K^+:Cs^+$ ratio for the oak roots was similar to that in the medium (Fig. 5.2). This result supports the possibility that K^+ and Cs^+ are not so discriminated during the root uptake. The value slightly increased in the shoot, indicating that oak has a mechanism to filter out K^+ during the root-to-shoot transport process.

5.5 Conclusions

The results of the experiment in hydroponically grown oak seedlings are consistent with precious findings in herbaceous plants in terms of the general theory that K^+ supply reduces Cs^+ uptake. However, there are noticeable differences in each aspect of the mechanism. It is suggested that oak root tissue uptakes K^+ and Cs^+ without apparent discrimination, and that the effect of K^+ supply in reducing Cs^+ accumulation is mainly due to ion competition. The property of the molecule functioning in K^+–Cs^+ uptake in oak root can be different from the known K^+–Cs^+ uptake transporters found in herbaceous plants.

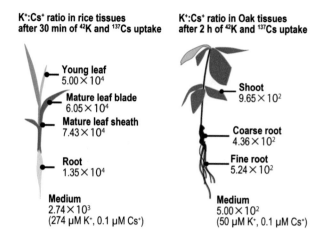

K⁺:Cs⁺ ratio in rice tissues
after 30 min of ⁴²K and ¹³⁷Cs uptake

Young leaf
5.00×10^4

Mature leaf blade
6.05×10^4

Mature leaf sheath
7.43×10^4

Root
1.35×10^4

Medium
2.74×10^3
(274 μM K⁺, 0.1 μM Cs⁺)

K⁺:Cs⁺ ratio in Oak tissues
after 2 h of ⁴²K and ¹³⁷Cs uptake

Shoot
9.65×10^2

Coarse root
4.36×10^2

Fine root
5.24×10^2

Medium
5.00×10^2
(50 μM K⁺, 0.1 μM Cs⁺)

Fig. 5.2 Comparison of ion selectivity on uptake and transport processes between rice and oak. The values for rice were cited from the precious report (Kobayashi et al. 2016) and those for oak seedlings were calculated using the values in Table 5.3

References

Epstein E, Rains DW, Elzam OE (1963) Resolution of dual mechanisms of potassium absorption by barley roots. Proc Natl Acad Sci U S A 49(5):684–692

Handley R, Overstreet R (1961) Effect of various cations upon absorption of carrier-free cesium. Plant Physiol 36(1):66–69

Kato N, Kihou N, Fujimura S, Ikeba M, Miyazaki N, Saito Y, Eguchi T, Itoh S (2015) Potassium fertilizer and other materials as countermeasures to reduce radiocesium levels in rice: results of urgent experiments in 2011 responding to the Fukushima Daiichi nuclear power plant accident. Soil Sci Plant Nutr 61(2):179–190

Kobayashi NI, Sugita R, Nobori T, Tanoi K, Nakanishi TM (2016) Tracer experiment using ⁴²K⁺ and ¹³⁷Cs⁺ revealed the different transport rates of potassium and caesium within rice roots. Funct Plant Biol 43(2):151–160

Kobayashi R, Kobayashi NI, Tanoi K, Masumori M, Tange T (2019) Potassium supply reduces cesium uptake in Konara oak not by an alteration of uptake mechanism, but by the uptake competition between the ions. J Environ Radioact 208:106032

Kubo K, Fujimura S, Kobayashi H, Ota T, Shinano T (2017) Effect of soil exchangeable potassium content on cesium absorption and partitioning in buckwheat grown in a radioactive cesium-contaminated field. Plant Product Sci 20(4):396–405

Nihei N, Tanoi K, Nakanishi TM (2018) Effect of different Cs concentrations on overall plant growth and Cs distribution in soybean. Plant Product Sci 21(1):26–31

Qi Z, Hampton CR, Shin R, Barkla BJ, White PJ, Schachtman DP (2008) The high affinity K⁺ transporter AtHAK5 plays a physiological role in planta at very low K⁺ concentrations and provides a caesium uptake pathway in *Arabidopsis*. J Exp Bot 59(3):595–607

Rubio F, Alemán F, Nieves-Cordones M, Martínez V (2010) Studies on *Arabidopsis athak5, atakt1* double mutants disclose the range of concentrations at which AtHAK5, AtAKT1 and unknown systems mediate K uptake. Physiol Plant 139(2):220–228

Zhu YG, Smolders E (2000) Plant uptake of radiocaesium: a review of mechanisms, regulation and application. J Exp Bot 51(351):1635–1645

Chapter 6
Candidates for Breeding Target Genes Related to Cesium Transport in Plants After the Fukushima Daiichi Nuclear Power Plant Accident

Keitaro Tanoi

6.1 Introduction

A large area in Eastern Japan was contaminated by radiocesium (^{137}Cs and ^{134}Cs) derived from the Tokyo Electric Power Company's Fukushima Daiichi Nuclear Power Plant (TEPCO-FDNPP) accident in 2011. At that time, there had been a lot of knowledge about Cs behavior in the environment (air, soil, water, etc.) and living things (plants, animals, microorganisms, etc.). Most of the knowledge had been obtained through research on the Chornobyl accident. Therefore, we, local agricultural researchers around Fukushima, searched for the Cs knowledge among the old journals to comprehend what would happen in food contamination because of the accident. As a result, we could obtain numerous meaningful and practical information about counter measurement in agriculture, for example, potassium amelioration on Cs accumulation in crops and contribution of soil minerals on Cs fixation in soil. In addition, there was a lot of data about Cs accumulation tendency among crops; for example, berries and mushrooms likely have higher Cs concentrations, and grains have much lower Cs.

Cs$^+$ transport mechanism in plants has been vigorously researched before the TEPCO-FDNPP accident in 2011. Previous researches revealed that the uptake mechanism of Cs$^+$ from soil to root was similar to that of K$^+$ (Collander 1941; Bange and Overstreet 1960; White and Broadley 2000). Consequently, reports on K$^+$ transporters mediating Cs$^+$ in plants increased (Kim et al. 1998; Qi et al. 2008). Especially HAK5, a high-affinity transport system belonging to the K$^+$ uptake permease transporter/high-affinity K$^+$ transporter/K$^+$ transporter (KUP/HAK/KT)

K. Tanoi (✉)

Graduate School of Agricultural and Life Sciences, The University of Tokyo, Tokyo, Japan

e-mail: uktanoi@g.ecc.u-tokyo.ac.jp

© The Author(s) 2023

T. M. Nakanishi, K. Tanoi (eds.), *Agricultural Implications of Fukushima Nuclear Accident (IV)*, https://doi.org/10.1007/978-981-19-9361-9_6

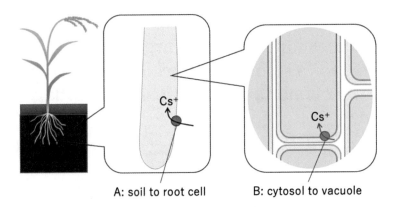

A: soil to root cell   B: cytosol to vacuole

Fig. 6.1 Cesium ion transport mechanism in plant roots. (**a**) Cesium transport from soil solution to root cell. For example, KUP/HAK/KT family and ATP-binding cassette (ABC) proteins, ABCG37 and ABCG33, are transporters mediating the Cs^+ transport process. (**b**) Cesium transport from cytosol to vacuole inside cells. For example, the SNARE protein, Sec22p/SEC22, was reported to be involved in the transport system

family, was found to be the main contributor mediating Cs^+ uptake from soil to root (Qi et al. 2008). Furthermore, in addition to the KUP/HAK/KT family, KAT1 and SKOR1 were shown to mediate Cs^+ by electrophysiology (see the review of Zhu and Smolders 2000; White and Broadley 2000). Other than potassium transporters, cyclic nucleotide (CN)-gated and calmodulin (CaM)-regulated channels (CNGCs) have been candidates to transport Cs^+. For example, AtCNGC10 might allow Cs^+ to cells when AtCNGC10 was expressed in *E. coli* (Li et al. 2005).

TEPCO-FDNPP accident has been promoting the research on every aspect of Cs^+, and the studies on the molecular mechanism of Cs^+ transport in plants have also progressed. In this chapter, I summarized the reports for 11 years, from 2011 to 2022, about transporters and transportation systems, which have possibilities to be employed for making low-Cs crops (Fig. 6.1).

6.2 Potassium Transporters, KUP/HAK/KT Family

The potassium transporter of *Arabidopsis*, HAK5, was already known as the primary contributor of Cs^+ uptake from soil to plant under low potassium conditions before the TEPCO-FDNPP accident, 2011 (Qi et al. 2008). However, the primary transporters mediating Cs^+ uptake in rice had been unknown. In 2017, OsHAK1 was identified to mediate Cs^+ uptake from soil to plants by reverse genetics (Nieves-Cordones et al. 2017). In the paper, OsHAK1 was firstly expressed in yeast and shown to mediate Cs^+ uptake in yeast. Then *OsHAK1*-edited mutant plants were produced using the CRISPR/Cas9 genome editing technique and examined Cs^+ uptake in hydroponic culture solution and soil derived from Fukushima. As a result,

Cs concentration in *OsHAK1*-edited mutant plants was drastically reduced compared with the wild-type plant when the plants grew on the low potassium soil. The results implied that OsHAK1 would be a breeding target for making low-cesium rice. However, *OsHAK1*-edited mutants have a penalty for obtaining potassium from low-K soil (the potassium concentration was about half that of wild-type plants), resulting in low biomass in *OsHAK1*-edited mutants under low potassium conditions (Nieves-Cordones et al. 2017). The strategy of simple knock-out OsHAK1 seems not to work. On the other hand, upgrading ion selectivity would be a potential strategy for making low-Cs rice. Cs^+ concentration in soil is so low (~10 μM) that the OsHAK1 selectivity of K^+ against Cs^+ has not been subjected to evolutionary pressure. Interestingly, a single mutation of amino acid residue (F130S) in the Arabidopsis HAK5 increases K^+ selectivity over Cs^+ (Alemán et al. 2014), suggesting room for improving K^+ selectivity over Cs^+ in other plants and crops. OsHAK1 was again identified as a causal gene of low-Cs rice in the same year by a forward genetic strategy using Japonica rice cv. Akitakomachi (Rai et al. 2017). In the paper, potassium concentration and biomass of *oshak1* were not changed compared with the wild-type. The results indicate that some cultivars could obtain enough potassium from other K^+ transport systems even though the cultivar lacks OsHAK1. It would be a valuable strategy to find rice cultivars with strong potassium absorption under low K condition and then mutate the OsHAK1.

OsSOS2 is not directly related to potassium or cesium transport, but mutation in OsSOS2 reduced Cs in rice grain (Ishikawa et al. 2017). Under Na^+ stress, OsSOS2 activates OsSOS1, which is a Na^+ efflux transporter. The expression levels of *OsHAK1, OsHAK5, OsAKT1, and OsHKT2;1* were about half in the OsSOS2 null mutant compared with WT under low-K (0.02 mM) Na existence (10 mM) condition. It was suggested that some of the four potassium transporters (OsHAK1, OsHAK5, OsAKT1, and OsHKT2;1) mediate Cs^+ uptake and transport to grains. Because the mutant did not have any penalty in growth, it would be an attractive plant to grow under radiocesium-contaminated land.

6.3   SNARE Sec22p/SEC22

The SNARE protein Sec22p/SEC22 functions for Cs accumulation in yeast and plants (Dräxl et al. 2013). The yeast mutant, sec22△, reduced Cs accumulation in the cell but did not change Rb and K concentration. The mathematical model predicted the Cs deposition in the vacuole. They also investigated the homologous gene, SEC22, in the model plant *Arabidopsis*. They found a *sec22* mutant that loses its expression in root, stem, and leaves but remains the expression in reproductive organs and young tissue. As a result, the *sec22* mutant reduced Cs concentration by half in leaves and 25% in roots. Even though the detailed mechanism of Sec22p/SEC22 for Cs^+ transport remains unknown, the SNARE protein would be a breeding target to produce low-Cs crops.

Retaining toxic elements in the vacuole of roots is one of the promising methods to reduce toxic element accumulation in grains. There is an excellent example of cadmium. OsHMA3 is a Cd^+ transporter that mediates Cd from cytosol to vacuole. OsHMA3-overexpression rice reduced Cd translocation from roots to shoots, resulting in 94–98% diminished Cd concentration in grains (Lu et al. 2019). In the case of Cs, there have been no reports about transporters specifically mediating Cs^+ from the cytosol to the vacuole in roots. It is unknown that the SNARE Sec22p/SEC22 protein can be used for increasing Cs concentration in the vacuole. Further studies about Cs^+ transport in the vacuole are needed.

6.4   ATP-Binding Cassette (ABC) Proteins, ABCG37 and ABCG33

The ATP binding cassette (ABC) transporters play essential roles in transporting various substances. Recently, ABC transporters are also involved in transporting minerals and metals; for example, arsenic (OsABCC1 and OsABCC2: Song et al. 2010), cadmium and mercury (OsABCC1 and OsABCC2: Park et al. 2012), cadmium (AtABCB25/AtATM3: Kim et al. 2006; AtABCG36/AtPDR8: Kim et al. 2007), lead (AtABCG4: Lee et al. 2005), and cesium (ABCG37 and ABCC33: Ashraf et al. 2021).

ABCG33 and ABCG37 have been shown to mediate Cs^+ in the wide range of Cs concentrations in the medium (0.4–10 mM). Importantly, ABCG33 and ABCG37 did not transport Rb^+ in yeast, suggesting that the ABCG33 and ABCG 37 can recognize the difference between Cs^+ and Rb^+. Furthermore, the plant of double knock-out of ABCG33 and ABCG37 showed reduced Cs^+ uptake compared with every single knock-out plant and wild type. Interestingly, the tendency of reduced Cs^+ uptake manner of the mutants was not changed under high and low K^+ conditions, indicating that potassium conditions do not influence the ABCG33 and ABCG37 function as Cs^+ transport. The Cs^+ uptake system that does not depend on K condition would help make crops of low-Cs contents under a standard-K field.

References

Alemán F et al (2014) The F130S point mutation in the Arabidopsis high-affinity K^+ transporter AtHAK5 increases K^+ over Na^+ and Cs^+ selectivity and confers Na^+ and Cs^+ tolerance to yeast under heterologous expression. Front Plant Sci 5:430

Ashraf MA et al (2021) ATP binding cassette proteins ABCG37 and ABCG33 function as potassium-independent cesium uptake carriers in Arabidopsis roots. Mol Plant 14:664–678

Bange GGJ, Overstreet R (1960) Some observations on absorption of cesium by excised barley roots. Plant Physiol 35:605–608

Collander R (1941) Selective absorption of cations by higher plants. Plant Physiol 16:691–720

Dräxl S et al (2013) Caesium accumulation in yeast and plants is selectively repressed by loss of the SNARE Sec22p/SEC22. Nat Commun 4:2092

Ishikawa S et al (2017) Low-cesium rice: mutation in OsSOS2 reduces radiocesium in rice grains. Sci Rep 7:2432

Kim EJ, Kwak JM, Uozumi N, Schroeder JI (1998) AtKUP1: an Arabidopsis gene encoding high-affinity potassium transport activity. Plant Cell 10:51–62

Kim D-Y et al (2006) AtATM3 is involved in heavy metal resistance in Arabidopsis. Plant Physiol 140:922–932

Kim D, Bovet L, Maeshima M, Martinoia E, Lee Y (2007) The ABC transporter AtPDR8 is a cadmium extrusion pump conferring heavy metal resistance. Plant J 50:207–218

Lee M, Lee K, Lee J, Noh EW, Lee Y (2005) AtPDR12 contributes to lead resistance in Arabidopsis. Plant Physiol 138:827–836

Li X, Borsics T, Harrington HM, Christopher DA (2005) Arabidopsis AtCNGC10 rescues potassium channel mutants of *E. coli*, yeast and Arabidopsis and is regulated by calcium/calmodulin and cyclic GMP in *E. coli*. Funct Plant Biol 32:643–653

Lu C et al (2019) Producing cadmium-free Indica rice by overexpressing OsHMA3. Environ Int 126:619–626

Nieves-Cordones M et al (2017) Production of low-Cs^+ rice plants by inactivation of the K^+ transporter OsHAK1 with the CRISPR-Cas system. Plant J 92:43–56

Park J et al (2012) The phytochelatin transporters AtABCC1 and AtABCC2 mediate tolerance to cadmium and mercury. Plant J 69:278–288

Qi Z et al (2008) The high affinity K^+ transporter AtHAK5 plays a physiological role in planta at very low K^+ concentrations and provides a caesium uptake pathway in Arabidopsis. J Exp Bot 59:595–607

Rai H et al (2017) Cesium uptake by rice roots largely depends upon a single gene, HAK1, which encodes a potassium transporter. Plant Cell Physiol 58:1486–1493

Song W-Y et al (2010) Arsenic tolerance in Arabidopsis is mediated by two ABCC-type phytochelatin transporters. Proc Natl Acad Sci 107:21187–21192

White PJ, Broadley MR (2000) Tansley Review No. 113. Mechanisms of caesium uptake by plants. New Phytol 147:241–256

Zhu YG, Smolders E (2000) Plant uptake of radiocaesium: a review of mechanisms, regulation and application. J Exp Bot 51:1635–1645

Chapter 7
Evaluation of the Absorption of Different Forms of Cesium from Soil

Makoto Furukawa, Shuichiro Yoshida, and Naoto Nihei

7.1 Introduction

The Great East Japan Earthquake that occurred in March 2011 triggered a tsunami that caused extensive damage. In addition to the damage caused by the tsunami, a nuclear accident occurred as a result of an accident at the Tokyo Electric Power Company's Fukushima Daiichi Nuclear Power Station (hereinafter referred to as "the nuclear accident"). To ensure safe agriculture in this region in the future, the form of radioactive cesium with a long half-life (half-life of 30.2 years, ^{137}Cs) that exists in the agricultural environment and how it is transferred to crops need to be elucidated.

Cesium in soil exists in two forms. One form is electrostatically adsorbed to the surface of mineral particles and can be easily dissolved, while the other form enters adsorption sites such as deep clay layers, binds stably to soil, and cannot be easily dissolved. The radioactive cesium that was released because of the nuclear accident gradually shifted from an easily soluble form to a less soluble form, reaching equilibrium at a certain rate (Fan et al. 2014).

The transfer factor (TF), which is the transfer ratio of radioactive cesium from soil to crops, is often used to evaluate the absorption of cesium (Cs) in soil by crops. However, currently used evaluation methods cannot determine the adsorption of Cs

M. Furukawa
PerkinElmer Japan Co., Ltd., Yokohama City, Kanagawa, Japan

S. Yoshida
Graduate School of Agricultural and Life Sciences, The University of Tokyo, Tokyo, Japan

N. Nihei (✉)
Faculty of Food and Agricultural Sciences, Fukushima University, Fukushima City, Fukushima, Japan
e-mail: nihei@agri.fuhukushima-u.ac.jp

© The Author(s) 2023
T. M. Nakanishi, K. Tanoi (eds.), *Agricultural Implications of Fukushima Nuclear Accident (IV)*, https://doi.org/10.1007/978-981-19-9361-9_7

from different soil fractions and cannot accurately evaluate the Cs absorption characteristics of different crop species.

In addition to the radioactive cesium (^{137}Cs) released from the Fukushima Daiichi Nuclear Power Plant, stable isotopes of cesium (^{133}Cs) exist in nature. Both ^{137}Cs and ^{133}Cs behave in the same manner, but the fractions present in the soil differ. In this study, we evaluated the absorption of Cs by different crops in different soil fractions by measuring the ^{133}Cs and ^{137}Cs concentrations in the soil and crops. We also evaluated the effect of potassium fertilization on cesium suppression in different soil fractions.

7.2 Cesium in Each Fraction of Soil

The soil in Fukushima Prefecture contaminated with radioactive cesium was classified into two fractions: the fraction of cesium with weak adsorption to soil, extracted using 1 M ammonium acetate and hydrogen peroxide (F1), and the fraction of cesium with strong adsorption to soil (F2), which was obtained by subtracting F1 from the cesium completely decomposed using hydrofluoric acid. ^{137}Cs in the cultivated crops and in the classified soil was measured using a germanium semiconductor detector and ^{133}Cs was evaluated using ICP-MS.

^{137}Cs abundances in the F1 and F2 fractions were approximately 2000 and 17,000 Bq/kg, 10.4% and 89.6% of the total ^{137}Cs, respectively. ^{133}Cs abundances in the F1 and F2 fractions were approximately 130 and 2670 µg/kg, 4.7% and 95.3% of the total ^{133}Cs, respectively. Differences were observed in the fractions of ^{137}Cs and ^{133}Cs. From soil samples collected in 2013, ^{137}Cs, which fell as a result of the nuclear accident, was more abundant than ^{133}Cs in the F1 fraction.

7.3 Intercrop Differences in ^{137}Cs and ^{133}Cs

Soybean, buckwheat, and rice were grown in pots with or without K fertilizer. Figure 7.1 shows the ^{137}Cs and ^{133}Cs contents [µg/kg] for each crop. When no K fertilizer was added, buckwheat absorbed the highest amount of ^{137}Cs per unit weight, whereas soybean and buckwheat absorbed similar amounts of ^{133}Cs per unit weight.

The inhibition of cesium absorption by K fertilization was observed for all crops. Compared with no K fertilization, the percentage reduction in Cs absorption was −31% ^{137}Cs and −52% ^{133}Cs for soybean, −32% ^{137}Cs, and −21% ^{133}C for buckwheat, and −33% ^{137}Cs and −32% ^{133}Cs for rice.

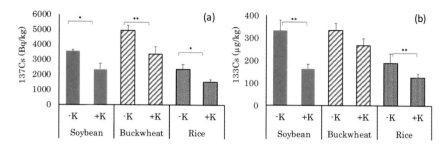

Fig. 7.1 Comparison of the effect of K fertilization on the suppression of (**a**) ^{137}Cs and (**b**) ^{133}Cs absorption for different crops. Significant at 5% (*) and 1% (**) assessed using t test among the same crops with/without K fertilization

7.4 Transfer Factor for Cs Present in Each Fraction

The Cs absorption from each fraction by the crops was then evaluated. When the transfer factor of Cs from the F1 fraction is TF_1 and the soil cesium concentration is C_1, and when the transfer factor of Cs from the F2 fraction is TF_2 and the soil cesium concentration is C_2, the cesium content in the crop (C_{crop}) can be expressed as follows:

$$C_1 \times TF_1 + C_2 \times TF_2 = C_{crop}. \tag{7.1}$$

This equation can be applied to both ^{133}Cs and ^{137}Cs.

$$^{133}C_1 \times TF_1 + {}^{133}C_2 \times TF_2 = {}^{133}C_{crop} \tag{7.2}$$

$$^{137}C_1 \times TF_1 + {}^{137}C_2 \times TF_2 = {}^{137}C_{crop}. \tag{7.3}$$

Because crops are not considered to absorb ^{133}Cs and ^{137}Cs separately, TF_1 and TF_2 in Eqs. (7.2) and (7.3) are the same for ^{133}Cs and ^{137}Cs. Therefore, by solving each TF from simultaneous Eqs. (7.2) and (7.3), the following equations are obtained:

$$TF_1 = \frac{{}^{137}C_2 \times {}^{133}C_{crop} - {}^{133}C_2 \times {}^{137}C_{crop}}{{}^{137}C_2 \times {}^{133}C_1 - {}^{133}C_2 \times {}^{137}C_1} \tag{7.4}$$

$$TF_2 = \frac{{}^{137}C_1 \times {}^{133}C_{crop} - {}^{133}C_1 \times {}^{137}C_{crop}}{{}^{137}C_1 \times {}^{133}C_2 - {}^{133}C_1 \times {}^{137}C_2}. \tag{7.5}$$

Figure 7.2 shows the transition coefficients calculated using Eqs. (7.4) and (7.5).

TF_1 tended to be larger in the absence of K fertilization, and buckwheat had the highest TF_1 value among the crops. With K fertilization, no significant differences were observed among soybeans, buckwheat, and rice. In contrast, TF_2 was less than

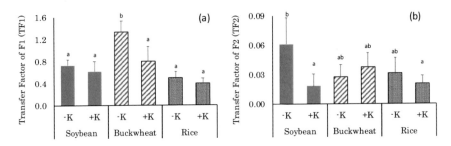

Fig. 7.2 Cs transfer coefficients for crops from (**a**) the F1 fraction (TF$_1$) and (**b**) the F2 fraction (TF$_2$). Significant differences according to the *t*-test are indicated by different letters

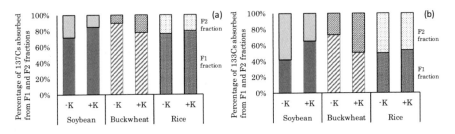

Fig. 7.3 Percentage of (**a**) ^{137}Cs and (**b**) ^{133}Cs absorbed from F1 and F2 fractions

one-tenth of TF$_1$, being the highest in soybeans in the absence of K fertilization. Conventional transfer factor evaluation could not determine which fraction of the soil absorbed Cs, but this method made it possible to determine the transfer from each fraction of soil.

The amount of cesium absorbed by the crop from each fraction was calculated from the transfer coefficient and the cesium concentration in each fraction. The percentage of Cs in the crop from each fraction is shown in Fig. 7.3. Without K fertilization, ^{137}Cs was absorbed mostly from the F$_1$ fraction by all three crops; ^{133}Cs was absorbed mostly from the F$_2$ and F$_1$ fractions by soybean and buckwheat, respectively, and from both F$_1$ and F$_2$ fractions by rice. Potassium fertilization decreased the percentages of both ^{137}Cs and ^{133}Cs absorbed from the F$_2$ fraction by soybean. In contrast, buckwheat presented reduced absorption from the F$_1$ fraction. In rice, the uptake of cesium by fraction did not change with or without potassium fertilization, indicating that different crops absorb cesium from different soil fractions.

7.5 K Fertilization Suppresses the Absorption of Each Cs Fraction

Furthermore, the difference in cesium concentration in the crop after K fertilization (C_{cropK}) was obtained by multiplying the TF from each fraction absorbed by the crop by the transfer inhibition factor k. The inhibition factor for fraction F1 (soil concentration is $C_{1'}$) is k_1, and that for fraction F2 (soil concentration is $C_{2'}$) is k_2.

$$^{133}C_1 \times (\mathrm{TF}_1 \times k_1) + {}^{133}C_2 \times (\mathrm{TF}_2 \times k_2) = {}^{133}C_{cropK} \tag{7.6}$$

$$^{137}C_1 \times (\mathrm{TF}_1 \times k_1) + {}^{137}C_2 \times (\mathrm{TF}_2 \times k_2) = {}^{137}C_{cropK}. \tag{7.7}$$

Therefore, the inhibition factors k_1 and k_2 from each fraction could be obtained by simultaneously solving Eqs. (7.8) and (7.9).

$$k_1 = \frac{{}^{137}C_2 \times {}^{133}C_{cropK} - {}^{133}C_2 \times {}^{137}C_{cropK}}{\mathrm{TF}_1 \left({}^{137}C_2 \times {}^{133}C_1 - {}^{133}C_2 \times {}^{137}C_1 \right)} \tag{7.8}$$

$$k_2 = \frac{{}^{137}C_1 \times {}^{133}C_{cropK} - {}^{133}C_1 \times {}^{137}C_{cropK}}{\mathrm{TF}_2 \left({}^{137}C_1 \times {}^{133}C_2 - {}^{133}C_1 \times {}^{137}C_2 \right)}. \tag{7.9}$$

It was determined that different crops absorb Cs differently from different soil fractions. The cesium absorbed by soybean without K fertilization was higher than that by buckwheat and rice, indicating that soybean could absorb more cesium that was strongly adsorbed to the soil. In soybean, $k_1 > k_2$, and the suppression of the transfer coefficient to the crop by K fertilization had a significant effect on the F2 fraction (Fig. 7.4). Thus, the results suggest that soybean absorbs a greater proportion of K from the F2 fraction in fields with a low K content. Buckwheat had a higher percentage and amount of Cs absorbed from the F1 fraction than soybean and rice, and actively absorbed Cs that was weakly adsorbed in the soil. K fertilization suppressed Cs absorption from the F1 fraction by buckwheat. Because buckwheat

Fig. 7.4 Results for the inhibition factor of K fertilization for each soil fraction

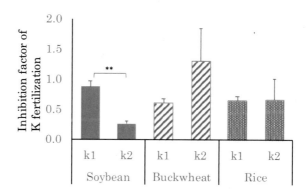

has $k_1 < k_2$, the Cs suppression effect of K fertilization also affected the absorption from the F1 fraction. Rice absorbed less Cs per unit weight than soybeans and buckwheat. Rice had $k_1 = k_2$, with no difference in absorbed fractions observed upon K fertilization in the cultivation test with this test soil, and Cs absorption was similar for both F1 and F2 fractions.

The sequential extraction results for the soils used in this study showed that the extraction rate of ^{137}Cs from the F1 fraction was higher than that of ^{133}Cs, indicating that the soil conditions facilitated the utilization of ^{137}Cs by crops. The difference in Cs uptake by each crop depends on which fraction of Cs is used among the various forms of Cs in the soil. This method that could calculate TF_1 and TF_2 could identify which fraction of Cs in the crop was derived from each fraction of soil. The results indicate that although the Cs in the crops tested in this study was mainly absorbed from the F1 fraction, most of the Cs was present in the F2 fraction of the soil, which was highly immobilized, and thus contributed a large proportion to the Cs absorption, even if TF_2 was small. The extent of absorption from the strongly immobilized F2 fraction of the soil was also determined for each crop. We also determined the suppression of Cs absorption by K fertilization. It has been shown that the transfer coefficient differs among crops, and that it is higher for soybean than for rice as a result of analysis of data after the nuclear power plant accident (IAEA 2020). We hope to clarify the differences in transfer coefficients among crops in the future by utilizing the method presented in this study.

However, this method could not determine the mechanism of absorption from the F2 fraction. It is necessary to clarify through future research whether crops directly absorb Cs from the F2 fraction; whether an equilibrium reaction occurs for Cs, transferring it from the F2 fraction because of a decrease in concentration in the F1 fraction, or a dissolution reaction induced by root-derived organic acids; and the relationship between this indicator and the F2 fraction. The relationship between the Cs concentration in the F1 fraction and this indicator should also be discussed.

References

Fan Q, Yamaguchi N, Tanaka M, Tsukada H, Takahashi Y (2014) Relationship between the adsorption species of cesium and radiocesium interception potential in soils and minerals: an EXAFS study. J Environ Radioact 138:92–100

IAEA (2020) Environmental transfer of radionuclides in Japan following the accident at the Fukushima Daiichi Nuclear Power Plant, TECDOC-1927. IAEA, Vienna

Chapter 8
Structure, Composition, and Physicochemical Properties of Radiocesium-Bearing Microparticles Emitted by the Fukushima Daiichi Nuclear Power Plant Accident

Taiga Okumura, Noriko Yamaguchi, and Toshihiro Kogure

8.1 Introduction

A significant amount of radionuclides was released into the environment during the accident at TEPCO's Fukushima Daiichi Nuclear Power Plant in March 2011, resulting in radioactive pollution in the surrounding area (Kinoshita et al. 2011). Among the released radionuclides, ^{137}Cs is a major source of the high air dose rate in the evacuation zone, because it was released in large quantities and has not yet decayed owing to its long half-life of approximately 30 years. The radiocesium released in a gaseous state during the accident fell to the ground with raindrops and was sorbed mainly to minerals in soil. For example, the Abukuma Mountains located west of the nuclear power plant are covered with weathered granitic soil called "masa" and most of the radiocesium deposited in this region was sorbed to partially vermiculitized biotite, which is abundant in masa (Mukai et al. 2014, 2016). Radiocesium is now strongly fixed to these mineral particles and is practically nonleachable by ion exchange (Mukai et al. 2018).

In addition, part of the released radiocesium was contained in microparticles that were formed in the damaged reactor and subsequently released into the environment. These particles, referred to as radiocesium-bearing microparticles (CsMPs) or "cesium balls," were first discovered in 2013 in aerosol filters collected in Tsukuba, 170 km southwest of the nuclear power plant (Adachi et al. 2013). CsMPs are almost

T. Okumura (✉) · T. Kogure
Department of Earth and Planetary Science, Graduate School of Science, The University of Tokyo, Tokyo, Japan
e-mail: okumura@eps.s.u-tokyo.ac.jp

N. Yamaguchi
Institute for Agro-Environmental Sciences, NARO, Tsukuba, Ibaraki, Japan

© The Author(s) 2023
T. M. Nakanishi, K. Tanoi (eds.), *Agricultural Implications of Fukushima Nuclear Accident (IV)*, https://doi.org/10.1007/978-981-19-9361-9_8

spherical and less than a few microns in diameter, and the radioactivity ratio of
$^{134}Cs/^{137}Cs$ suggests formation in reactor Units 2 or 3 (Satou et al. 2018). Furthermore, CsMPs have been suggested to originate from Unit 2 according to their
constituent elements (IRID 2018). Although the CsMP formation process has not
yet been fully elucidated, they may have been formed by condensation from the gas
phase, because most possess a spherical shape. Extremely high specific radioactivity
(radioactivity per volume) distinguishes CsMPs from the abovementioned
radiocesium-sorbing minerals (hereafter abbreviated as CsSMs). CsMPs were
transported and deposited over a wide area, extending into the Kanto region, because
of their small size (Abe et al. 2021). CsMPs are unique in that they have never been
reported in past nuclear power plant accidents. Therefore, understanding the structure, composition, and physicochemical properties of CsMPs, such as their thermal
and dissolution characteristics, is an urgent issue. In this chapter, recent CsMP
characterization efforts using electron microscopy are detailed. A lingering issue is
the need to establish a method for quantitatively estimating CsMP abundance in
environmental samples to accurately understand their contribution to radioactive
pollution and influence on the environment and human health. Here, we propose an
approach to resolve this issue based on the different dissolution properties of CsSMs
and CsMPs in acidic solution, as they cannot be discriminated only by γ-ray
measurements. Meanwhile, larger radiocesium-bearing particles of more than several tens of microns were found within 20 km of the nuclear power plant. These
particles, referred to as "type-B" particles and probably released from Unit 1, have
distinct characteristics from CsMPs (type-A) and have been characterized in other
studies (e.g., Satou et al. 2018; Igarashi et al. 2019).

8.2 CsMP Structure and Composition

The CsMPs investigated in the studies described below were collected from sources
including contaminated soil, plant tissues, bird feathers, and aerosols, but most
abundantly from nonwoven fabric cloth laid on an agricultural field in the central
part of Fukushima Prefecture. The procedure for collecting CsMPs has been
described previously (Yamaguchi et al. 2016). Because the CsMPs are less than a
few microns in size, their structure has been investigated mainly by scanning
electron microscopy (SEM) and conventional and scanning transmission electron
microscopy (TEM/STEM) with energy-dispersive X-ray spectroscopy (EDS). In
particular, improvement in the detection efficiency of X-rays in EDS has enabled
the determination of element distribution in very small CsMPs (Lechner et al. 2001).
Electron-transparent thin films for the TEM/STEM analyses of CsMPs were prepared using a focused ion beam (FIB) system with microsampling equipment.

8.2.1 Silicate Glass Matrix of CsMPs

Following the discovery of CsMPs, synchrotron X-ray microbeam diffraction suggested that the CsMPs were amorphous (Abe et al. 2014). Subsequently, TEM/STEM observations and elemental analyses by EDS showed that CsMPs are mainly composed of silicate glass containing Cl, K, Fe, Zn, Rb, Sn, and Cs as the major constituents (Fig. 8.1) (Yamaguchi et al. 2016). More recently, X-ray absorption near-edge structure (XANES) analyses using scanning transmission X-ray microscopy (STXM) and STEM-EDS analysis revealed that Na is also a major constituent of CsMPs, and its atomic percentage is larger than that of the other alkali elements (K, Rb, and Cs) (Okumura et al. 2020b). The trace presence of Mn, Mo, Te, Ba, and U has been reported using synchrotron-beam X-ray fluorescence and high-energy-resolution EDS with microcalorimetry (Abe et al. 2014; Kogure et al. 2016). However, it is not clear whether these trace elements are contained within the silicate glass matrix or in the nanoparticles present inside the CsMPs, as discussed below.

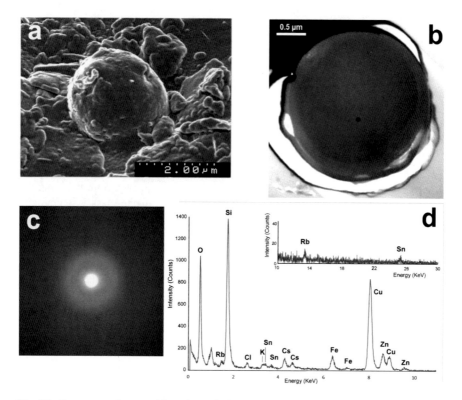

Fig. 8.1 Structure and composition of a typical CsMP. (**a**) Secondary-electron SEM image of a CsMP. (**b**) Bright-field TEM image and (**c**) electron diffraction pattern of a thin-sectioned CsMP prepared using FIB. (**d**) EDS spectrum acquired from a CsMP. Cu was emitted from the supporting grid. (Adapted from Yamaguchi et al. 2016)

Meanwhile, B was not detected in the CsMPs even using electron energy-loss spectroscopy, which is quite sensitive to the detection of light elements, implying that B_4C control rods might have created a eutectic alloy with stainless steel cladding (Okumura et al. 2019a). It should be noted that Al and Ca were not present inside the CsMPs, although some studies have detected them using EDS analysis with SEM (Satou et al. 2016, 2018; Higaki et al. 2017; Miura et al. 2018). These elements may have been derived from fine mineral particles attached to the surface of CsMPs.

Elemental mapping of CsMPs by STEM-EDS revealed that not all elements were uniformly distributed within the particles (Fig. 8.2) (Kogure et al. 2016; Okumura et al. 2019a). Cesium showed the most prominent inhomogeneous distribution, with a concentration higher near the surface and lower in the central of the particle. The nonuniform distribution of Cs can be explained by the inward diffusion of gaseous Cs into the suspended silicate melt balls during condensation. Following the inhomogeneous distribution of Cs, Cl showed a similar pattern. In contrast, other alkali elements (such as K and Rb) tend to concentrate in the inner part. In addition, a few CsMPs exhibited high concentrations of Fe and Zn near the surface. These nonuniform distributions of the constituent elements are frequently observed in CsMPs, but their concentration gradients vary greatly between particles (Fig. 8.3) (Kogure et al. 2016). Therefore, it can be assumed that the environment in which the CsMPs were formed inside the damaged reactor is not uniform. In some CsMPs, a high concentration of Sn was observed around particle surfaces, likely due to the precipitation of insoluble SnO_2 during CsMP dissolution after their release into the environment (Rai et al. 2011).

The atmosphere in which the CsMPs were formed in the damaged reactor could also be estimated from the valence state of the constituent elements. XANES analysis was performed using STXM to investigate the valence state of Fe (Fig. 8.4) (Okumura et al. 2020b). Inside the CsMPs, Fe^{2+} was dominant, while Fe^{3+} increased toward the upper region. The thin specimens for STXM were prepared using FIB and the sample thinned toward the upper region, implying that the increased Fe^{3+} is an artifact. In addition, the Fe in silicate glass was gradually oxidized by X-ray irradiation during the STXM experiments. Considering this change induced by the measurement, the Fe state in CsMPs was probably almost Fe^{2+} at the time of formation, suggesting that the reactor was in a significantly reducing atmosphere. The valence state map of Fe in Fig. 8.4 also shows that Fe was highly oxidized near the particle surface likely due to oxidation in the environment.

8.2.2 Nanoparticles in CsMPs

Using TEM/STEM observations, various types of nanoparticles have been found inside CsMPs, including oxides with a spinel structure containing Cr and Fe (Fig. 8.5) (Okumura et al. 2019a). Generally, metallic ions with a small ionic radius and high valence, such as Cr^{3+}, hardly dissolve in silicate glass, forming or segregating oxide nanoparticles in the CsMPs. In addition to oxides, sulfides such as

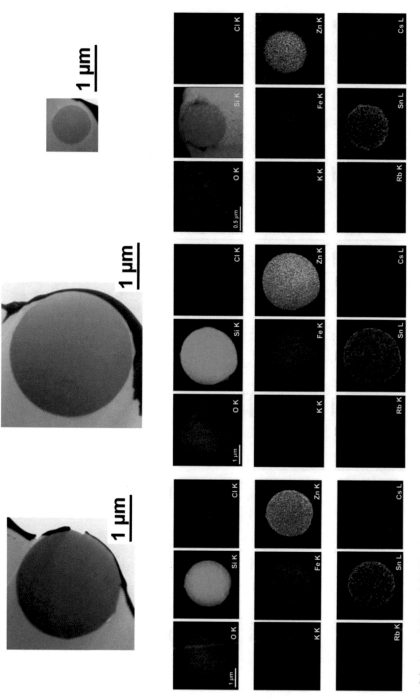

Fig. 8.2 (**upper**) Bright-field TEM images of three CsMPs and (**lower**) their elemental maps acquired using STEM-EDS. (Adapted from Kogure et al. 2016)

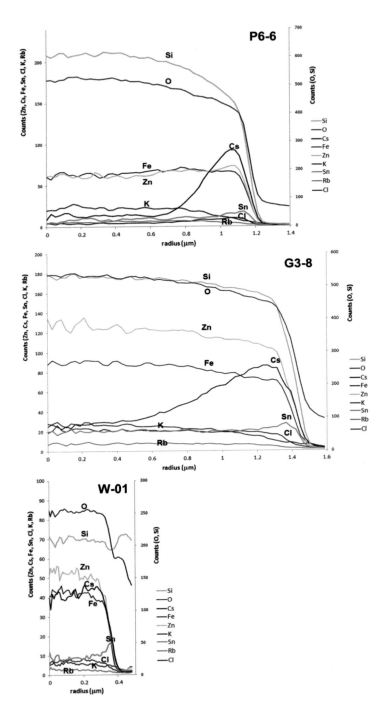

Fig. 8.3 Radial distribution of constituent elements in the three CsMPs shown in Fig. 8.2. (Reprinted from Kogure et al. 2016)

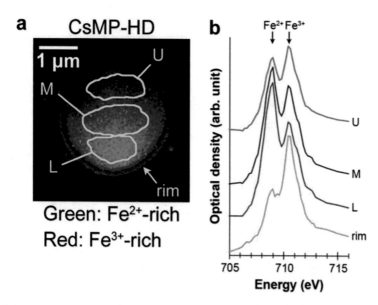

Fig. 8.4 (**a**) Fe valence state map acquired from a CsMP using STXM. (**b**) XANES spectra acquired from the L, M, and U areas in (**a**). The spectrum from the rim region is also shown. (Reprinted from Okumura et al. 2020b)

$Cu_{1.8}S$, Ag_2S, and MoS_2, and tellurides including Ag_2Te, have been found (Fig. 8.5) (Yamaguchi et al. 2016; Okumura et al. 2019a). It should be noted that the specimens observed were thin films processed by FIB; therefore, only a small portion of a single CsMP particle was examined using TEM/STEM. Nonetheless, these nanoparticles have been frequently detected, indicating that they are quite common in CsMPs.

8.2.3 CsMPs with Irregular Forms and Different Compositions

As described above, most CsMPs are almost spherical and their glass matrices exhibit similar compositions with several notable exceptions (Yamaguchi et al. 2017). Two representative examples of these anomalies are presented in Fig. 8.6. The left one, with an intensely foamed appearance, was found in 2013 in atmospheric dust. The specific radioactivity and $^{134}Cs/^{137}Cs$ ratio were comparable to those of the normal CsMPs, and TEM/STEM analysis indicated that crystalline materials were not present inside the particles, and the oxide glass was mainly composed of silicate and chromium-oxide. The former contained Al and Cs with an irregular distribution, while Fe, Zn, and Sn were present in both glass types but were more concentrated in the chromium-oxide glass. The CsMP on the right side of

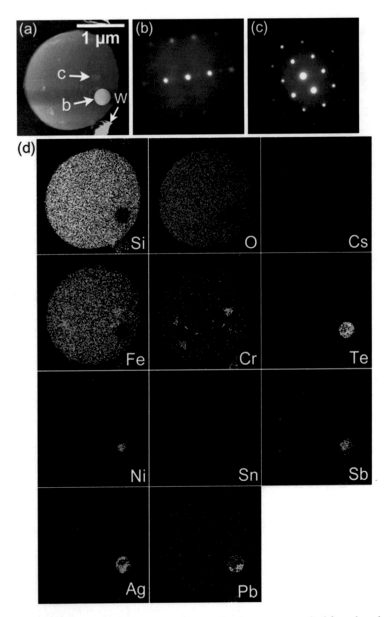

Fig. 8.5 (a) STEM image of a CsMP. (b) Electron diffraction pattern acquired from the spherical inclusion indicated by arrow "b" in (a), corresponding to hessite (Ag_2Te) observed along <113>. (c) Electron diffraction pattern acquired from the inclusion indicated by arrow "c" in (a), corresponding to a spinel structure observed along <114>. (d) Elemental maps of the CsMP. (Adapted from Okumura et al. 2019a)

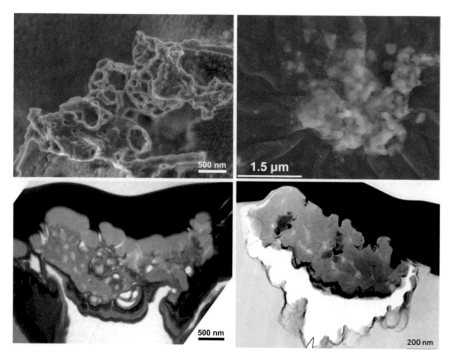

Fig. 8.6 Two examples of CsMPs with nonspherical forms. (**Left**) CsMP with an intensely foamed appearance. (**Right**) CsMP with an aggregate of fine particles. The upper images are the secondary-electron SEM images and the bottom shows bright-field TEM images of these particles. Note that the right CsMP is half-buried in glue. (Adapted from Yamaguchi et al. 2017)

Fig. 8.6 was recovered from plant tissue collected in Fukushima Prefecture in 2015. These images show an aggregate of finer particulates 100–200 nm in size composed of silicate glass containing Sn and Cs but without Fe and Zn. Iron was present as individual fine particles of hematite and magnetite in these aggregates.

The origin of these irregular and uncommon CsMPs must be completely different from that of the spherical counterparts. For instance, the foamed particle likely experienced rapid depressurization due to hydrogen explosions that occurred in Units 1 and 3. Considering its ^{134}Cs/^{137}Cs ratio, it is suspected that this particle originated from Unit 3.

8.3 Physicochemical Properties of CsMPs

As mentioned above, radioactive particulate materials like CsMPs have not been reported until the Fukushima nuclear accident and their physicochemical properties remain largely unknown. Therefore, several groups have investigated the properties of CsMPs to predict their dynamics and fates in the environment.

8.3.1 Thermal Properties of CsMPs

Radiation-contaminated waste is often incinerated to reduce its volume and is expected to contain CsMPs. However, the effects of incineration on CsMPs is unknown so isolated CsMPs were held on a platinum plate and heated in air at various temperatures to better understand their thermal properties (Okumura et al. 2018). The radioactivity of the CsMPs gradually decreased starting at 600 °C and was almost lost when the temperature reached 1000 °C. The size and spherical morphology of the CsMPs remained almost unchanged after heating, but the alkali elements (K, Rb, and Cs) and Cl were lost (Fig. 8.7). This indicates that these elements, including radiocesium, diffused away from the CsMPs and were volatilized upon heating, resulting in decreased radioactivity. Iron, zinc, and tin, which were originally dissolved in the glass matrix, formed various oxides including $ZnFe_2O_4$, Zn_2SiO_4, and SnO_2 inside the CsMPs. These elements diffused inside the CsMPs via heating and crystallized during cooling. In contrast, when the CsMPs were heated together with the soil, radiocesium released from the CsMPs was sorbed to the surrounding soil. Thus, these studies indicate that CsMPs lose their high specific radioactivity when radiation-contaminated waste is incinerated at sufficiently high temperatures.

8.3.2 Dissolution Properties of CsMPs

CsMPs have been recognized as insoluble particles, because their shapes and radioactivity did not change when immersed in water (Adachi et al. 2013). However, CsMPs are mainly composed of silicate glass, which can slowly dissolve in aqueous

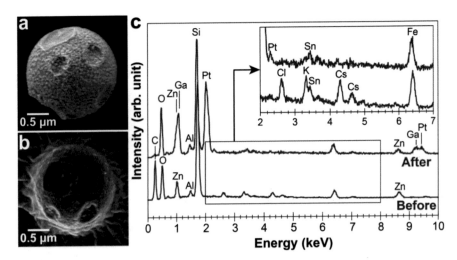

Fig. 8.7 (**a**) SEM image of a CsMP before heating. (**b**) SEM image of the CsMP after heating at 900 °C. (**c**) EDS spectra acquired from the CsMP before and after heating. (Adapted from Okumura et al. 2018)

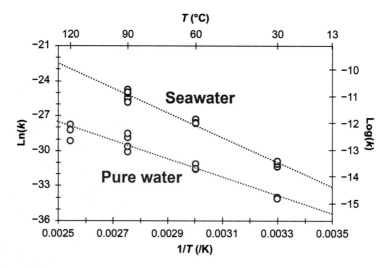

Fig. 8.8 Arrhenius plot of the logarithm of k (rate of decrease in the CsMP radii) as a function of the reciprocal temperature $1/T$ for pure water and seawater. (Reprinted from Okumura et al. 2019b)

solution. Therefore, CsMP solubility should be further investigated to predict the environmental fates of these particles.

Dissolution experiments were performed using CsMPs in pure water and artificial seawater at several temperatures and the dissolution progress was monitored by the decrease in the ^{137}Cs radioactivity originating from the CsMPs, allowing the dissolution rate and activation energy to be estimated (Okumura et al. 2019b). The dissolution rate of the CsMPs in seawater was approximately one order of magnitude higher than that in pure water (Fig. 8.8). Assuming that a 1 μm radius CsMP is immersed in seawater at 13 °C (the approximate annual mean temperature in Fukushima City), it will completely dissolve within 10 years. Furthermore, the CsMPs were observed using SEM, TEM, and STEM after the dissolution experiments. The shapes of the CsMPs dissolved in pure water were considerably altered, suggesting that the decreased radioactivity was caused not by elution of ^{137}Cs from the silicate glass through ion-exchange, but by dissolution of the glass itself. Tin and iron originally in the glass matrix formed oxide nanoparticles on the CsMP surfaces owing to their low solubility (Fig. 8.9). This is similar to observations of CsMPs collected in Fukushima Prefecture (Yamaguchi et al. 2017), indicating that CsMP dissolution actually occurs in the environment. For CsMPs dissolved in seawater, a crust composed of secondary Mg- and Fe-rich minerals was formed on the surface and glass matrix dissolution proceeded inside the crust (Fig. 8.9).

Subsequently, further dissolution experiments in solutions of various compositions and pH values were conducted at 60 °C. The dissolution behavior of CsMPs was largely comparable to that of silica-rich glass under these variable conditions (Fig. 8.10). In neutral and basic solutions, dissolution was accelerated by alkali ions such as Na$^+$, which played a catalytic role. In contrast, dissolution in acid was slow,

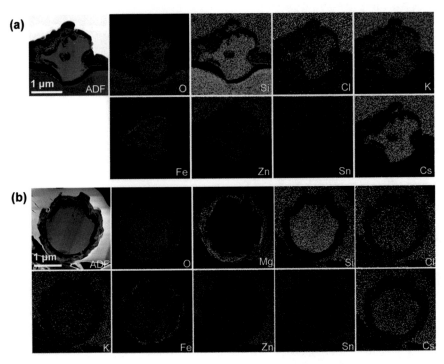

Fig. 8.9 STEM images and corresponding element maps of CsMPs after the dissolution experiments in pure water (**a**) and artificial seawater (**b**). (Adapted from Okumura et al. 2019b)

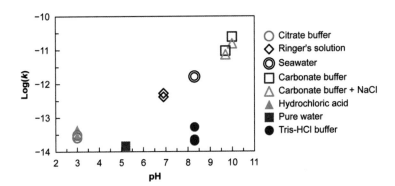

Fig. 8.10 Dependence of dissolution rates (k) on pH at 60 °C, where k (m/s) is the decrease rate of the CsMP radii in the presence (open symbols) and absence (closed symbols) of Na ions

even in the presence of alkali ions. In addition, the dissolution rate in Ringer's solution at 37 °C was determined to be 1.00 ± 0.37 μm/year, implying that CsMPs with radii of approximately 1 μm will dissolve completely within a few years at most if they are inhaled.

8.4 Discrimination of CsMPs in Contaminated Samples

CsMPs have spread widely in eastern Japan, with several atmospheric plumes emitted from damaged reactors on March 14–15, 2011 (Tsuruta et al. 2015; Nakajima et al. 2017). Hence, their deposition areas and abundance were controlled by environmental factors such as the wind transporting the plumes, rainfall, and landscapes. Furthermore, CsMPs deposited in fields, forests, and cities have been exposed to environmental factors that may have caused movement during approximately a decade after the accident. However, abundance and distribution of CsMPs in the environment remains largely unknown. Autoradiography using an imaging plate (IP) is a common method to identify CsMPs as "radioactive particles," but a large number of CsSMs are also present in environmental samples. Recently, a method to distinguish CsMPs from CsSMs using autoradiography was developed based on the assumption that radioactivity per individual particle is higher in CsMPs than in CsSMs (Ikehara et al. 2018). However, CsSMs with higher radioactivity than that of CsMPs have been reported (Okumura et al. 2019c), indicating that this distinction method could be inaccurate because of the overlapping radioactivity distributions between CsMPs and CsSMs.

As an alternative approach, a discrimination method based on the difference in response to acidic solutions between CsMPs and CsSMs was developed (Okumura et al. 2020a). The dissolution experiments of the CsMPs mentioned above indicated that the dissolution rate of CsMPs in acidic solutions is very slow, with only mild reductions of radioactivity. In contrast, CsSMs treated with acidic solutions under the same conditions completely lost their radioactivity (Okumura et al. 2020a). Radiocesium sorbed or fixed on the surface of the CsSMs was almost completely released, because acidic solutions dissolve the surface atomic layers. Furthermore, contaminated samples (nonwoven fabric cloth and wheat leaves) were immersed in 1 mM HCl at 90 °C for 24 h, and then in 100 mM HCl at 90 °C for 24 h. The samples were exposed to IP after each treatment (Fig. 8.11) and the IP read-out image before treatment showed several hot spots and uniform luminescence intensity throughout the cloth. However, uniform luminescence was almost completely extinguished by immersion in 1 mM HCl, likely due to the release of radiocesium from the CsSMs. Meanwhile, almost all hot spots remained even after immersion in 100 mM HCl, which could be attributed to CsMPs, because these hot spots would have been extinguished if they originated from CsSMs. In addition to the discrimination of CsMPs using IP autoradiography, their contribution to the total radioactivity of contaminated materials can be estimated quantitatively using γ-ray measurements. When the contaminated samples were immersed in acidic solutions and radioactivity periodically measured, the decrease rate slowed after the initial rapid drop (Fig. 8.12). The initial drop was ascribed to CsSM dissolution and the contribution of CsMPs to the total radioactivity was easily estimated by measuring the residual radioactivity after the acid treatment.

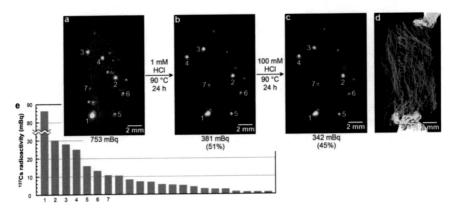

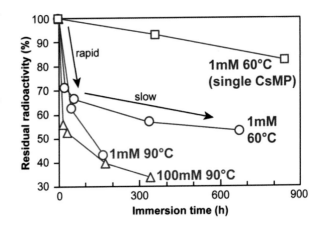

Fig. 8.11 IP read-out images (**a–c**) and optical micrograph (**d**) of a fragment of nonwoven fabric cloth. (**a**) Before acid treatment. (**b**) After immersion in 1 mM hydrochloric acid. (**c, d**) After immersion in 100 mM hydrochloric acid. The ^{137}Cs radioactivity corresponding to the entire area of the IP read-out images is noted below each image and the residual radioactivity after immersion is noted in brackets. (**e**) ^{137}Cs radioactivity determined from the hot spots in the IP read-out image shown in **b**. (Reprinted from Okumura et al. 2020a)

Fig. 8.12 Residual ratio of ^{137}Cs radioactivity of nonwoven fabric cloth fragments (circles and triangles) and single CsMP (squares) as a function of immersion time. They were immersed in 1 or 100 mM hydrochloric acid at 60 or 90 °C. (Reprinted from Okumura et al. 2020a)

8.5 Concluding Remarks

Knowledge of the composition and structure of CsMPs has increased considerably since their discovery in 2013 and it is generally agreed that CsMPs were formed in Unit 2. Fuel debris is recovered and investigated to understand the current conditions of the damaged reactors and guide their decommissioning processes. Debris investigations may provide further information regarding the CsMP formation processes and accident progression. The physicochemical properties of CsMPs, including their thermal and dissolution behaviors, have also been characterized to some extent, which will contribute significantly to the prediction of their dynamics in the environment and influence on human health. However, information regarding the abundance and distribution of CsMPs in Fukushima Prefecture and Kanto area is necessary to fully understand radioactive pollution by ^{137}Cs. For this, an approach to estimate the contribution of CsMPs to the radioactivity of environmental specimens was developed. In future research, this method should be improved to be more efficient and practical for the recovery of Fukushima and its surrounding regions.

Acknowledgements We are grateful to Prof. Y. Takahashi (University of Tokyo) for supporting the STXM measurements. This work was supported by the Japan Society for the Promotion of Science Grants-in-Aid for Scientific Research (19H01145, 20K19954), a research contract with the Japan Atomic Energy Agency for Fukushima Environmental Recovery, the University of Tokyo Advanced Characterization Nanotechnology Platform in the Nanotechnology Platform Project sponsored by the Ministry of Education, Culture, Sports, Science and Technology, Japan, and the Ministry of Agriculture, Forestry, and Fisheries contract research "Radioactivity surveys in Japan."

References

Abe Y, Iizawa Y, Terada Y et al (2014) Detection of uranium and chemical state analysis of individual radioactive microparticles emitted from the Fukushima nuclear accident using multiple synchrotron radiation X-ray analyses. Anal Chem 86:8521–8525

Abe Y, Onozaki S, Nakai I et al (2021) Widespread distribution of radiocesium-bearing microparticles over the greater Kanto Region resulting from the Fukushima nuclear accident. Prog Earth Planet Sci 8:13

Adachi K, Kajino M, Zaizen Y, Igarashi Y (2013) Emission of spherical cesium-bearing particles from an early stage of the Fukushima nuclear accident. Sci Rep 3:2554

Higaki S, Kurihara Y, Yoshida H et al (2017) Discovery of non-spherical heterogeneous radiocesium-bearing particles not derived from Unit 1 of the Fukushima Dai-ichi Nuclear Power Plant, in residences 5 years after the accident. J Environ Radioact 177:65–70

Igarashi Y, Kogure T, Kurihara Y et al (2019) A review of Cs-bearing microparticles in the environment emitted by the Fukushima Dai-ichi Nuclear Power Plant accident. J Environ Radioact 205:101–118

Ikehara R, Suetake M, Komiya T et al (2018) Novel method of quantifying radioactive cesium-rich microparticles (CsMPs) in the environment from the Fukushima Daiichi Nuclear Power Plant. Environ Sci Technol 52:6390–6398

IRID (2018) Upgrading of the comprehensive identification of conditions inside reactor: accomplishment report for FY2017

Kinoshita N, Sueki K, Sasa K et al (2011) Assessment of individual radionuclide distributions from the Fukushima nuclear accident covering central-East Japan. Proc Natl Acad Sci 108:19526–19529

Kogure T, Yamaguchi N, Segawa H et al (2016) Constituent elements and their distribution in the radioactive Cs-bearing silicate glass microparticles released from Fukushima nuclear plant. Microscopy 65:451–459

Lechner P, Fiorini C, Hartmann R et al (2001) Silicon drift detectors for high count rate X-ray spectroscopy at room temperature. Nucl Instrum Methods Phys Res Sect A Accel Spectr Detect Assoc Equip 458:281–287

Miura H, Kurihara Y, Sakaguchi A et al (2018) Discovery of radiocesium-bearing microparticles in river water and their influence on the solid-water distribution coefficient (K_d) of radiocesium in the Kuchibuto River in Fukushima. Geochem J 52:145–154

Mukai H, Hatta T, Kitazawa H et al (2014) Speciation of radioactive soil particles in the Fukushima contaminated area by IP autoradiography and microanalyses. Environ Sci Technol 48:13053–13059

Mukai H, Motai S, Yaita T, Kogure T (2016) Identification of the actual cesium-adsorbing materials in the contaminated Fukushima soil. Appl Clay Sci 121:188–193

Mukai H, Tamura K, Kikuchi R et al (2018) Cesium desorption behavior of weathered biotite in Fukushima considering the actual radioactive contamination level of soils. J Environ Radioact 190:81–88

Nakajima T, Misawa S, Morino Y et al (2017) Model depiction of the atmospheric flows of radioactive cesium emitted from the Fukushima Daiichi Nuclear Power Station accident. Prog Earth Planet Sci 4:2

Okumura T, Yamaguchi N, Dohi T et al (2018) Loss of radioactivity in radiocesium-bearing microparticles emitted from the Fukushima Dai-ichi nuclear power plant by heating. Sci Rep 8:9707

Okumura T, Yamaguchi N, Dohi T et al (2019a) Inner structure and inclusions in radiocesium-bearing microparticles emitted in the Fukushima Daiichi Nuclear Power Plant accident. Microscopy 68:234–242

Okumura T, Yamaguchi N, Dohi T et al (2019b) Dissolution behaviour of radiocaesium-bearing microparticles released from the Fukushima nuclear plant. Sci Rep 9:3520

Okumura T, Yamaguchi N, Kogure T (2019c) Finding radiocesium-bearing microparticles more minute than previously reported, emitted by the Fukushima nuclear accident. Chem Lett 48:1336–1338

Okumura T, Yamaguchi N, Kogure T (2020a) Distinction between radiocesium (RCs)-bearing microparticles and RCs-sorbing minerals derived from the Fukushima nuclear accident using acid treatment. Chem Lett 49:1294–1297

Okumura T, Yamaguchi N, Suga H et al (2020b) Reactor environment during the Fukushima nuclear accident inferred from radiocaesium-bearing microparticles. Sci Rep 10:1352

Rai D, Yui M, Schaef HT, Kitamura A (2011) Thermodynamic model for $SnO_2(cr)$ and $SnO_2(am)$ solubility in the aqueous Na^+–H^+–OH^-–Cl^-–H_2O system. J Solut Chem 40:1155–1172

Satou Y, Sueki K, Sasa K et al (2016) First successful isolation of radioactive particles from soil near the Fukushima Daiichi Nuclear Power Plant. Anthropocene 14:71–76

Satou Y, Sueki K, Sasa K et al (2018) Analysis of two forms of radioactive particles emitted during the early stages of the Fukushima Dai-ichi Nuclear Power Station accident. Geochem J 52:1–7

Tsuruta H, Oura Y, Ebihara M et al (2015) First retrieval of hourly atmospheric radionuclides just after the Fukushima accident by analyzing filter-tapes of operational air pollution monitoring stations. Sci Rep 4:6717

Yamaguchi N, Mitome M, Akiyama-Hasegawa K et al (2016) Internal structure of cesium-bearing radioactive microparticles released from Fukushima nuclear power plant. Sci Rep 6:20548

Yamaguchi N, Kogure T, Mukai H et al (2017) Structures of radioactive Cs-bearing microparticles in non-spherical forms collected in Fukushima. Geochem J 51:1–14

Chapter 9
Verification of Effects on Crops and Surrounding Environment in Agriculture Using Radioactively Contaminated Grass Silage Compost Made by Aerobic Ultrahigh Temperature Fermentation

Takahiro Yoshii, Yoshihiro Furui, Tairo Oshima, and Noboru Manabe

9.1 Compost Derived from Grass Silage Contaminated with Radioactive Cesium

The Fukushima Daiichi Nuclear Power Plant (Tokyo Electric Power Co. Ltd.) accident in 11 March, 2011, released enormous amounts of radionuclides, causing radioactive contamination of many plants, animals, soil, rivers, oceans, and other environments. More than 10 years have passed since that accident, but grass silage contaminated with Cs-134 (^{134}Cs) and Cs-137 (^{137}Cs) was left untouched in pastures and farmlands immediately after the accident. While some of the low-containing grass silage (less than 400 Bq/kg) has been plowed into the soil, the highly contaminated grass silage (approximately 8000 Bq/kg or less) is still left untreated.

In our previous studies, we have attempted to utilize microbial methods of organic waste reduction to convert highly contaminated grass silage into compost (Manabe et al. 2016; Yoshii et al. 2019) through aerobic ultrahigh temperature (more than 110 °C) compost fermentation techniques (Oshima and Moriya 2008; Moriya et al. 2011). The contaminated grass of about 2700 Bq/kg was converted to 297 Bq/kg of

T. Yoshii · Y. Furui
Institute of Environmental Microbiology, Kyowa Kako Co. Ltd., Tokyo, Japan

T. Oshima
Institute of Environmental Microbiology, Kyowa Kako Co. Ltd., Tokyo, Japan

Faculty of Bioscience and Biotechnology, Tokyo Institute of Technology, Tokyo, Japan

N. Manabe (✉)
Faculty of Agriculture, The University of Tokyo, Tokyo, Japan

Faculty of Human Sciences, Osaka International University, Osaka, Japan
e-mail: n-manabe@oiu.jp

© The Author(s) 2023
T. M. Nakanishi, K. Tanoi (eds.), *Agricultural Implications of Fukushima Nuclear Accident (IV)*, https://doi.org/10.1007/978-981-19-9361-9_9

compost. The compost made by aerobic ultrahigh temperature fermentation was applied to crops grown in the laboratory, but no radioactive cesium was transferred to the plants.

In the present study, we investigated the transfer of radioactive cesium to pasture grasses and its effects on the environment, including soil, space, and rivers, when compost is used on actual pastures.

9.2   Pasture Cultivation Using Compost Derived from Radioactively Contaminated Grass Silage

The experimental farmland in Kurihara-city, Miyagi-prefecture, Japan, located 150 km north of the Fukushima Daiichi Nuclear Power Plant, averaged 320 Bq/kg of radioactive cesium in June 2017 after radioactive decontamination (Fig. 9.1). This farmland originally was used for rice cultivation but was converted to pasture after the Fukushima Daiichi Nuclear Power Plant accident. The 1200 m^2 farmland was divided into 6 plots of 200 m^2 each, and 4 of these plots were applied with 2, 5, 10, and 30 kg of radioactively contaminated grass silage compost (297 Bq/kg, made by aerobic ultra-high temperature fermentation techniques) per m^2 using manual spreader or excavator, and the remaining 2 plots were fertilized with chemical fertilizers or cattle manure as control (Fig. 9.2). After those plots were plowed, pasture Orchard grass (*Dactylis glomerata*) was sown and grown for about 2 months starting in June 2017. No fertilizer or pesticide was applied. In the first 3 weeks after sowing, the 30 kg/m^2 compost application caused the growth disturbance in the

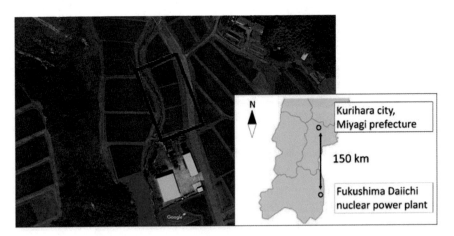

Fig. 9.1 The testing site (1200 m^2 pastureland, average 320 Bq/kg) in Kurihara-city, Miyagi-prefecture, Japan (150 km away from Fukushima Daiichi nuclear power plant) https://earth.google.com/web/@38.86985053,140.98021213,135.46996306a,211.35840595d,35y,358.53244018h,0t,0r

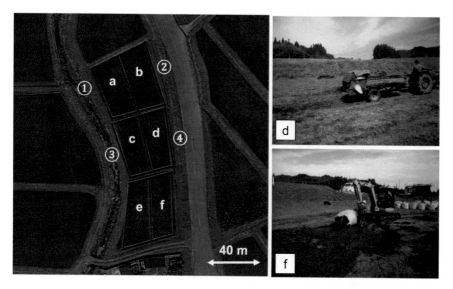

Fig. 9.2 Overview of test plots. Control plots were fertilized with chemical fertilizers (**a**) or cattle manure (**b**). The amounts of fertilizer applied were 2 kg (**c**), 5 kg (**d**), 10 kg (**e**), and 30 kg (**f**) per m^2 of radioactively contaminated grass compost (297 Bq/kg). Compost application by manure spreader (**b–d**) or excavator (**e**, **f**). Location of measurement of air dose rate (①–④) and river water radioactivity (① upstream, ③ outlet of drainage) during the cultivation experiment

Fig. 9.3 The pastures at 2 months after radioactively contaminated grass compost (made by aerobic ultra-high temperature fermentation) application. (**a–f**) Soil pH and exchangeable potassium increased with increasing compost usage. Grass growth was best in plot (**e**) where 10 kg/m^2 of grass compost was applied

pasture, but the pasture recovered after that. The other test plots showed good growth throughout the test period (Fig. 9.3). Soil pH and exchangeable potassium increased with increasing compost application. Sufficient potassium in the soil is thought to

Table 9.1 Radioactive cesium concentrations in soil before and after application of contaminated grass compost made by aerobic ultrahigh temperature fermentation

Date (2017)		Soil ^{134}Cs + ^{137}Cs concentrations (Bq/kg—dry weight)					
		a	b	c	d	e	f
26 April	Pretest	523	523	519	519	348	348
23 June	After fertilization	479	432	315	420	239	348
9 August	49 days after seeding	428	436	514	408	328	409

Table 9.2 The pH, radiocesium, and total nitrogen concentrations in river water upstream and the outlet of the test plots during the application of contaminated grass compost made by aerobic ultrahigh temperature fermentation

Date (2017)		① Upstream			② Outlet of drainage		
		^{134}Cs + ^{137}Cs (Bq/kg)	Total N (mg/L)	pH	^{134}Cs + ^{137}Cs (Bq/kg)	Total N (mg/L)	pH
9 June	Fertilization and plowing	<1	0.194	7.36	<1	0.134	7.30
23 June	Seeding (0 day)	<1	0.527	6.83	<1	0.485	6.87
14 July	23 days after seeding	<1	0.152	7.14	<1	0.139	7.10
28 July	37 days after seeding	<1	0.143	7.20	<1	0.139	7.29
9 August	49 days after seeding	<1	0.291	7.12	<1	0.299	7.03
28 August	68 days after seeding	<1	0.175	7.62	<1	0.100	7.55

compete with crop absorption of radiocesium (^{134}Cs + ^{137}Cs). The pastures were cut at 37, 49, and 68 days after sowing at 5 cm above the ground surface. The concentrations of radioactive cesium in the grasses were measured with a germanium semiconductor detector (Mirion Technologies Canberra, Tokyo, Japan). As a result, radioactive cesium levels in the leaves and stems of the grasses were not detected in any of the test plots (total levels of ^{134}Cs and ^{137}Cs were less than 20 Bq/kg). Genetic analyses of the cut grasses by polymerase chain reaction targeting the 5.8S ribosomal ITS region were performed and confirmed to be Japanese millet (*Echinochloa esculenta*) or Orchard grass. These weeds such as Japanese millet were thought to be species that were originally established in the test plots.

Each soil radioactivity did not change significantly before and after compost application (Table 9.1). Air dose rates at 1 m from the ground in four corners around the test sites before and after application of the radioactively contaminated grass silage compost were measured using a scintillation instrument (PA1000, Horiba, Kyoto, Japan). The results showed that the air dose rates before and after compost application ranged from 0.06 to 0.12 μSv per hour, with no significant changes due to the application of radioactively contaminated silage compost (Table 9.2).

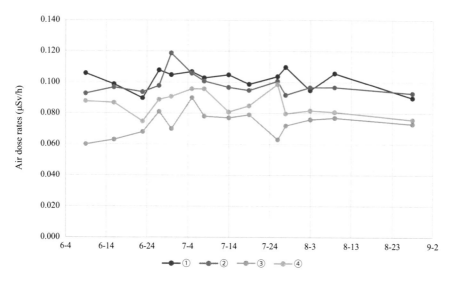

Fig. 9.4 Air dose rate around the test plots in 2017 (see Fig. 9.2) during pasture cultivation test using grass compost contaminated with radioactive cesium

Water samples were collected from upstream of the river adjacent to the test site and directly below the test site, just six times. No radioactive cesium was detected in any of the river waters. There was also no significant change in the total nitrogen content of river water before and after application of radiocesium contaminated grass silage compost (Fig. 9.4).

9.3 Conclusion

The pastures were grown with 2, 5, 10, and 30 kg of compost produced from contaminated grass silage applied per square meter, and the grass and weeds were undetectable for radioactivity. As the amount of contaminated grass compost applied increased, the pH in the soil increased and the amount of potassium also increased. Therefore, it may be expected to have a soil improvement effect to inhibit the absorption of radioactive cesium. No effect was observed on air dose rate or river radioactivity in the surrounding area where contaminated grass compost was applied. The results of this study confirm that crop cultivation using contaminated grass compost is safe in open field cultivation.

References

Manabe N, Takahashi T, Piao C, Li J, Tanoi K, Nakanishi T (2016) Adverse effects of radiocesium on the promotion of sustainable circular agriculture including livestock due to the Fukushima Daiichi nuclear power plant accident. In: Nakanishi TM, Tanoi K (eds) Agricultural implications of the Fukushima nuclear accident. Springer-Nature, Heidelberg, pp 91–98. https://doi.org/10.1007/978-4-431-55828-6_8

Moriya T, Hikota T, Yumoto I, Ito T, Terui Y, Yamagishi A, Oshima T (2011) *Calditerricola satsumensis* gen. Nov., sp. nov. and *Calditerricola yamamurae* sp. nov., extreme thermophiles isolated from a high-temperature compost. Int J Syst Evol Microbiol 61:631–636

Oshima T, Moriya T (2008) A preliminary analysis of microbial and biochemical properties of high temperature compost. Ann N Y Acad Sci 1125:338–344

Yoshii T, Oshima T, Matsui S, Manabe N (2019) A composting system to decompose radiocesium contaminated baled grass silage. In: Nakanishi T, O'Brien M, Tanoi K (eds) Agricultural implications of the Fukushima nuclear accident (III). Springer-Nature, Heidelberg, pp 51–58. https://doi.org/10.1007/978-981-13-3218-0_6

Chapter 10
Transport of ^{137}Cs into Fruits After External Deposition onto Japanese Persimmon Trees

Mamoru Sato

10.1 Introduction

The Fukushima Daiichi Nuclear Power Plant (FDNPP) is in Fukushima Prefecture, which is a major area of deciduous fruit production in Japan (Ministry of Agriculture, Forestry and Fisheries 2022). During the FDNPP accident, there was a large release of radiocesium (RCs) into the environment especially in the period 12–14 March 2011.

The transfer factor via soil ($CR_{f\text{-}s}$) is one of the parameters used to quantify the extent of transfer of RCs from soil into the edible parts of crops, defined as an activity concentration ratio on a weight basis. Prior to the Fukushima accident, several studies have been conducted to determine the $CR_{f\text{-}s}$ for fruit trees (IAEA (International Atomic Energy Agency) 2020). Another most commonly used parameter to quantify transfer is the aggregated transfer factor (T_{ag}), which is defined as the ratio of RCs activity concentration in edible parts (Bq kg^{-1}, FW or DW) to RCs deposition amount per unit area (Bq m^{-2}). T_{ag} does not necessarily imply that the soil is the source of contamination, as it quantifies the contribution via all migration processes leading to the radioactivity concentration of the fruit. However, there have been no previous reports of T_{ag} quantifying migration when the source is bark deposition.

At the time of the FDNPP accident, most deciduous fruit trees, except Japanese apricot [*Prunus mume* (Sieb.) Sieb. et Zucc.], were in the dormant stage prior to bud burst. Since most deciduous fruit trees had not developed leaves, only the bark of the plant was directly contaminated. Of the released radionuclides, ^{137}Cs and ^{134}Cs were the most important contributors to plant contamination.

M. Sato (✉)
Faculty of Food and Agricultural Sciences, Fukushima University, Fukushima, Fukushima, Japan
e-mail: florpomofores2105-1804@coral.plala.or.jp

© The Author(s) 2023
T. M. Nakanishi, K. Tanoi (eds.), *Agricultural Implications of Fukushima Nuclear Accident (IV)*, https://doi.org/10.1007/978-981-19-9361-9_10

Initial studies in May 2011 reported elevated RCs activity concentrations in leaves and young fruits of peach, of 449 Bq kg^{-1} FW and 157 Bq kg^{-1} FW, respectively, at the Fruit Tree Research Center, Fukushima Agricultural Technology Center (FTRC), approximately 65 km northwest of the FDNPP. Absorption of RCs from the roots was initially unlikely as deposited RCs was deposited onto and retained in the upper soil layers of orchard soil where there are few tree roots. Therefore, the data indicated that RCs attached to the surface of the bark may have contributed to fruit contamination (Sato et al. 2019a). Before the FDNPP accident, a small study by Katana et al. (1988) reported that RCs migrate directly into tree tissues from contaminated bark. After the FDNPP accident, several studies also found that ^{137}Cs deposited on the aboveground part of deciduous trees, including fruit trees, had migrated into the tree mainly via bark (Tagami et al. 2012; Takata 2013; Kuroda et al. 2013; Sato et al. 2014; Aoki et al. 2017). Consistent with these observations, bark-washing of Japanese persimmon and peach was effective in decreasing the ^{137}Cs activity concentration ($[^{137}$Cs]) in fruit in the subsequent year (2012) (Sato et al. 2015a).

Furthermore, the $[^{137}$Cs] in fruits of the Japanese persimmon, which did not have bark washed, was significantly higher than that of trees with washed bark at least for 7 years after the FDNPP accident (Sato et al. 2019b). Decontamination by external washing resulted in a reduction in RCs activity concentration of 29.1% in fruits and 33.2% in leaves in the Japanese persimmon (Sato et al. 2015a). These data indicated that residual ^{137}Cs on the bark continued to move to other parts of the unwashed tree for several years after ^{137}Cs was deposited. However, prior to this study, there had been no reports that have quantitatively verified the transfer of RCs from externally contaminated bark of dormant branches to fruits.

Dissolved ^{137}Cs (termed here as dis-^{137}Cs solution) was detected in stemflow more than 4 years after the FDNPP accident (Sato et al. 2017a, b, 2021). An alternative possible pathway for continuing contamination of fruit of the transfer was via stemflow of ^{137}Cs from external bark to leaves, which had developed after the FDNPP accident. The observation raised the need to quantify the relative importance of the externally deposited RCs compared with that from contaminated soil uptake over several years after deposition. The leaf-to-fruit transfer rate of RCs has been previously examined in the UK, France, Italy, and Germany after the Chernobyl accident. The reported leaf-to-fruit transfer of ^{134}Cs expressed as fraction of the applied or intercepted activity has been highly variable at 1.8–9.6% in grapes (Zehnder et al. 1995; Carini et al. 1996, 1999; Carini and Lombi 1997) and 1–47% in apples (Katana et al. 1988; Carini et al. 1999; Pröhl et al. 2003). The transport of RCs to fruits via leaves in these studies was affected by the following factors: stage of contamination (seasonality), precipitation intensity, age of branch and tree, cultivar and tree species, distance between fruit and the contaminated leaf, dilution effect from growth, and yield.

The translocation of assimilate in fruit trees depends on the relationship between source (leaves) and sink (fruits) (Warren 1972). RCs is an analogue of K and, therefore, behaves in a similar manner in plants. The direction of translocation and distribution of K with various assimilates such as organic acids and sugars (Van

Goor and Van Lune 1980; Possner and Kliewer 1985; Lang 1983; Westwood 1993) is controlled by the sink strength defined by Warren (1972) as: "sink strength = sink size × sink activity," which is represented by the growth rate of fruit (Faust 1989). Pröhl et al. (2003) found that the transfer to fruits via leaves was higher in apples after spraying ^{137}Cs solution during the fruit growing stage than at the flowering stage. Sink activity in fruit increases after fruit growing stage, so the transfer of ^{137}Cs is probably also affected by the relationship of source and sink between fruits and leaves.

Fruit load affects the sink-source relationship, and is often modified by the horticultural practice of fruit thinning that adjusts the fruit load using the leaf-fruit ratio as a criterion. Horii et al. (2019) revealed that there was no clear relationship between the leaf-fruit ratio and the [^{137}Cs] in the fruit during the defoliation study. However, there have been no previous studies that have examined the [^{137}Cs] in fruit due to fruit thinning.

Japanese persimmon has been cultivated and consumed for a long time in Japan. Ampo-gaki is a product of astringent persimmon fruit semidried in the sun or in the open air and is an important processed foodstuff produced largely in the northern part of Fukushima Prefecture. The production volume of ampo-gaki in 2010 was the highest recorded in Japan of 1423 t before the FDNPP accident (Ministry of Agriculture, Forestry and Fisheries 2010). Because drying fruit can lead to higher [^{137}Cs], ampo-gaki production has been under voluntary suspension in specific areas designated by Fukushima Prefecture because of the likelihood that [^{137}Cs] will exceed designated standard limits in Japan. Since 2013, an inspection system for [^{137}Cs] in ampo-gaki has been established, and production has resumed in areas designated by Fukushima Prefecture. Japanese persimmon has been, and remains, the most important economically damaged fruit product produced in Fukushima Prefecture after the FDNPP accident.

Japanese persimmon fruits are true fruits with enlarged ovaries. The flower buds are a mixture of the immature shoot structure and flower parts, and the fruit-bearing part is the fruiting mother shoot that grew the previous year. After sprouting, the growing fruiting shoots bear fruit at several leaf axils. Several fruiting shoots grow from a fruiting mother shoot, and several fruiting mother shoots grow from a lateral branch. A group of fruiting mother shoots is formed mainly on the lateral branches that are 2–5 years of age. There is a relatively large fruit load produced on the apical fruiting mother shoot that develops from the apical bud of the 2-year-old lateral branch. In an experiment conducted after the Chernobyl accident by Katana et al. (1988), the positional relationship between fruit and contaminated leaves within the same branch was not considered. It was reported that the [^{137}Cs] in fruits of Japanese persimmon varies widely even within the same lateral branch, but the cause of this variation was not clarified (Horii et al. 2019; Sato et al. 2021).

The surface of the peel of Japanese persimmon fruit lacks aerenchyma, and the respiration of fruit is carried out through a calyx. The calyx of the Japanese persimmon is a homologous organ comprising of leaves and has plenty of stomata and spongy tissues (Watanabe and Hoshi 1981). Since RCs is transported to pulp via the calyx in the Japanese persimmon (Sekizawa et al. 2016a, b), evaluation of the transfer rate of RCs to fruit via both the leaf and calyx is needed.

The aims of this work were to (1) quantify the transfer rate of ^{137}Cs applied to leaves and calyx into fruit; (2) examine effects of leaf position and fruit load on the transport of ^{137}Cs into fruit; and (3) quantify the transfer rates and aggregated transfer factors of ^{137}Cs applied to apical fruiting mother shoot and 2-year-old lateral branch into fruits before bud burst.

10.2 Materials and Methods

An outline of experiments conducted from 2016 to 2021 is shown in Table 10.1.

10.2.1 Preparation of ^{137}Cs Solution

For Experiment 1 (EXP. 1), moss samples of *Thuidium cymbifolium* (Dozy & Molk.) Dozy & Molk., collected from the Yamakiya area in the town of Kawamata approximately 50 km northwest of the FDNPP in May 2015, were used for preparing a ^{137}Cs solution. After boiling approximately 200 g of moss with 1.5 L of distilled water for 15 min, the moss solution was filtrated through a pulp sheet (JK Wiper, NIPPON PAPER CRECIA CO., Ltd., Tokyo) laid onto a stainless-steel net, and centrifuged at 3500 rpm. The supernatant was filtered through a 0.45 μm membrane filter (cellulose ester mixed type, ADNANTEC A045A047A, Toyo Roshi Kaisha, Ltd., Tokyo) under suction. The ^{137}Cs solution was reduced in volume to approximately 350 mL by heating and then subjected to high-pressure sterilization treatment in an autoclave (SX-700, TOMY SEIKO CO., Ltd., Tokyo) at 120 °C for 20 min to obtain a final solution containing 2.2 Bq mL^{-1} of dis-^{137}Cs. For EXP. 2 and EXP. 3, moss samples of *Plagiothecium nemorale* (Brid.) Z. Iwats.; *P. roseanum* Schimp. in the Nagadoro area of Iitate village located about 40 km northwest from the FDNPP and of *Thuidium cymbifolium* (Dozy & Molk.) Dozy & Molk. in the same site as EXP. 1 were collected in July 2016 and May 2017. The moss was used to prepare the experimental dis-^{137}Cs solution as described above producing 1.8 Bq mL^{-1} of dis-^{137}Cs solution. For EXP. 4, moss samples of *Pylaisiadelpha tenuirostris* (Buch et Schimp.) W.R. Buch. in the Nagadoro area of Iitate village were collected in May 2017. The moss was used to prepare the experimental dis-^{137}Cs solution as described above producing 7.0 for 2020 and 7.5 for 2021 Bq mL^{-1} of dis-^{137}Cs solution. In Experiments 1, 2, and 3, epiphytic moss growing on rock was used, whereas epiphytic moss on bark was used in Experiment 4.

Table 10.1 Outline of the experimental plan and the results related to the TR_f and the $T_{ag}^{f\text{-}b}$ into mature fruit

Experiment no./year	Cultivar	Comparative treatment: Application Stage	Applied organ	Applied position	Replicates (shoot/branch)	Application days	Amount of applied ^{137}Cs (Bq per replicate)	TR_f^a (%)	$T_{ag}^{f\text{-}b} \times 10^{-4}$ (m^2 kg^{-1} DW)
1/2016	Hachiya	Young fruit	Calyx	Front side of a calyx	3	24 June–1 July	12.3	14.9 ± 4.7	
		Fruit growing	Calyx	Front side of a calyx	4	1–10 August	21.1	10.3 ± 4.6	
		Young fruit	Leaf	Adjacent to fruit on a short fruiting shoot	2	24 June–1 July	25.0	10.2 ± 3.2	
		Fruit growing	Leaf	Adjacent to fruit on a short fruiting shoot	3	1–10 August	15.9	16.4 ± 5.0	
2/2017	Hiratanenashi	Young fruit	Leaf	Adjacent to fruit on a long fruiting shoot	4	26-30 June	12.5	8.4 ± 2.6	
			Leaf	Middle part in a long fruiting shoot	4			2.7 ± 1.2	
			Leaf	Top part in a long fruiting shoot	4			2.0 ± 0.8	
			Leaf	Adjacent to fruit on a short fruiting shoot	6			8.8 ± 4.3	
3/2018	Hiratanenashi	Fruit growing	Leaf	All leaves on the central fruiting shoot on the fruiting mother shoot with thinning or with no thinning	4	July 31–2 August	7.6	17.1 ± 6.1	
								10.0 ± 1.8	
				Adjacent to fruit on a short fruiting shoot	3			12.8 ± 4.7	
4/2020	Hiratanenashi	Dormancy	Shoot	Front side of an apical fruiting mother shoot with thinning	8	3 and 6 April	13.3	1.3 ± 0.6	15.2 ± 6.9
			Branch	Front side of a 2-year-old lateral branch with thinning	8			0.89 ± 0.35	9.1 ± 8.8
4/2021			Shoot	Front side of an apical fruiting mother shoot with no thinning	4	7 April	13.1	3.2 ± 1.8	8.0 ± 4.5

^{a}Arithmetic mean ± standard deviation

10.2.2   Measurement of the Amount of ^{137}Cs Applied to Various Organs

The [^{137}Cs] in several treated organs (i.e., calyx, leaf, shoot) collected immediately after treatment was measured, and the retained ratio (*RR*) was calculated using Eq. (10.1). The amount of ^{137}Cs (Bq) in individual treated organ was calculated using Eq. (10.2):

$$RR = w_{so} \times C_{so} \times \left(s_g \times v_s \times C_s\right)^{-1}, \tag{10.1}$$

where w_{so} (kg DW) is total weight of the contaminated organs, C_{so} (Bq kg^{-1} DW) is the [^{137}Cs] in the contaminated organs, s_g ($= 1.00$ g mL^{-1}) is the specific gravity of the applied dis-^{137}Cs solution, v_s (mL) is the volume of the applied dis-^{137}Cs solution, C_s (Bq g^{-1}) is the [^{137}Cs] in the dis-^{137}Cs solution.

$$AS_{io} = RR \times s_g \times v_s \times C_s, \tag{10.2}$$

where AS_{io} is the amount of ^{137}Cs in individual contaminated organs (Bq organ^{-1}), *RR* is the retained ratio, v_s (mL) is the volume of the applied dis-^{137}Cs solution, C_s (Bq g^{-1}) is the [^{137}Cs] in the dis-^{137}Cs solution.

10.2.3   Transport into Japanese Persimmon Fruits of ^{137}Cs Applied on Calyx and Leaves (EXP. 1)

10.2.3.1   Plant Material and Treatments

The experiments were conducted using two 12-year-old "Hachiya" Japanese persimmon (*Diospyros kaki* Thunb.) trees planted at the FTRC. Fruit bearing shoots less than about 20 cm in length were selected for the ^{137}Cs application treatment. The experiments were performed both at the young and the growing stage of fruits. In 2016, the fruit young stage was June 24–July 1 (23–31 days after full blossom) and the fruit growing stage was August 1–15 (61–75 days after full blossom). To compare the amount of ^{137}Cs transported to fruit from the calyx or from the leaves, the dis-^{137}Cs solution was applied onto either the front side of the calyx or onto two adjacent leaves using a bottle with a plastic brush (Fig. 10.1). A cellulose cloth was placed onto the calyx to prevent contamination by ^{137}Cs during application onto the adjacent leaves, which was repeated 1–3 times per day. After the treatment, the applied solution was left to dry naturally, and the fruits were then wrapped within a net bag. To calculate the *RR*, the [^{137}Cs] was measured in three sets of contaminated leaves collected immediately after each treatment and in 5 and 6 contaminated calyxes on June 27 and July 1, respectively, for the fruit young treatment, and in 3 contaminated calyxes on August 1 for the fruit growing stage treatment.

Fig. 10.1 Procedure to contaminate the plants by applying the dis-^{137}Cs solution to the calyx or leaves of the Japanese persimmon "Hachiya" in EXP. 1 conducted in 2016. A sheet of cellulose cloth was placed beneath the calyx to prevent adhesion of the treatment solution to the calyx during the leaf contamination process. The dis-^{137}Cs solution was applied on the surface of the front side of calyx and onto two leaves adjacent to the fruit using a bottle with a plastic brush

There were losses of contaminated fruit, commonly called June drop, caused by factors such as pollination and lack of sunshine (Sobashima and Takagi 1968; Kitajima et al. 1987, 1990; Chauhan 2014). During the period from the start of July to mid-August, the young unripe fruits, which had dropped into the net, were collected (on July 4 and 11) and the ^{137}Cs was measured for both treatments (calyx and leaf contamination).

10.2.3.2 Sample Preparation

Contaminated leaves and mature fruits were collected on October 26 and the ^{137}Cs in leaves and fruit was measured. Fruit was separated into calyx, peel, and pulp. On July 27 and October 31, respectively, 7 and 6 uncontaminated fruits were also sampled to measure the ^{137}Cs for control of the young fruit and the mature fruit. The ^{137}Cs was measured for each part of the combined control fruit sample in a similar manner to that for contaminated fruits.

10.2.4 Effects of Leaf Position on the Transport into Fruit of [137]Cs Applied on Leaves (EXP. 2)

10.2.4.1 Plant Material and Treatments

The experiments were conducted using three 13-year-old "Hiratanenashi" Japanese persimmon trees planted at FTRC. The effects of the distance between leaves and fruit on leaf to fruit transport of ^{137}Cs were examined. An outline of the treatments is shown in Table 10.1 and Fig 10.2. From June 26th to June 30th during the juvenile period, 20.3 Bq of ^{137}Cs was applied to 2 leaves as explained below. Twelve fruiting shoots that were longer than 20 cm (long-fruiting shoots) and bearing one fruit were selected. Two leaves were contaminated from each of the three positions: adjacent to

Fig. 10.2 The treatment of Japanese persimmon "Hiratanenashi" cultivar in EXP. 2 conducted in 2017. The dis-^{137}Cs solution was applied on the surface of the front side of two leaves from the adjacent to fruit, middle and top part of the long fruiting shoots (>25 cm length) and two leaves adjacent to fruit on the short fruiting shoot (<15 cm length) using a bottle with a plastic brush

the fruit, at the middle node, and at the apical node. Similarly, six short-fruiting shoots (less than 20 cm in length) were selected, where two leaves adjacent to the fruit were contaminated. The ^{137}Cs solution was applied over the surface of the front side of leaves using the same method as in EXP. 1. The [^{137}Cs] in six treated leaves collected within a few seconds after treatment was measured, and the RR was calculated.

10.2.4.2   Sample Preparation

Samples of the Japanese persimmon were collected on October 18. Each replicate of fruit and contaminated leaves were prepared individually for each treatment position group. Leaves that had not been contaminated from each long fruiting shoot were collected and subdivided into the upper and the lower leaf from the node of treated leaves.

10.2.5   Effects of Fruit Load on the Transport into Fruit of [137]Cs Applied on Leaves (EXP. 3)

10.2.5.1   Plant Material and Treatments

The experiments were conducted using a 15-year-old "Hiratanenashi" Japanese persimmon tree planted at FTRC. Eight fruiting mother shoots with 5 consecutive fruiting shoots longer than 20 cm (long fruiting shoots) bearing one fruit were selected, and comparative treatments by fruit load were applied as follows:

Four "No fruit thinning" group where each of five consecutive fruiting shoots bears one fruit (Fig. 10.3a).

Four "Fruit thinning" group where fruits on the adjacent shoots of the central one were picked off just before the contamination treatment took place on July 31 (Fig. 10.3b).

Three short fruiting-shoots (less than 20 cm) developed on other lateral branches were selected to compare with the above-mentioned long fruiting shoot groups. Replicates were made on four fruiting mother shoots.

Dis-^{137}Cs solution was applied to all leaf surfaces of the central (among five) fruiting shoots and of the short fruiting shoots during the period from July 31 to August 2. The total volume of applied dis-^{137}Cs solution was 12 mL. The [^{137}Cs] in leaves of three treated fruiting shoots collected within a few seconds after treatment was measured, and the *RR* was calculated.

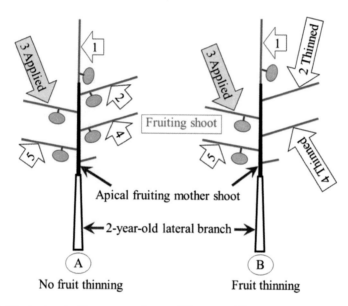

Fig. 10.3 The treatment of Japanese persimmon "Hiratanenashi" cultivar in EXP. 3 conducted in 2018. Eight fruiting mother shoots with 5 "No fruit thinning" shoots longer than 20 cm (long fruiting shoots) that bear one fruit were selected. Prior to applying dis-^{137}Cs solution, the comparative group was set as follows. "No fruit thinning" group (**a**): One fruit was set on each of the five "No fruit thinning" shoots. "Fruit thinning" group (**b**): Fruits on the upper and lower adjacent shoots of the central shoot among the five consecutive shoots were thinned. Dis-^{137}Cs solution was applied to all leaf surfaces of the central fruiting shoots among five fruiting shoots and the short fruiting shoots

10.2.5.2 Sample Preparation

Fruits, leaves, and wood with bark of fruiting shoots and fruiting mother shoots were collected on October 22. Same organs were also collected from uncontaminated lateral branches in the same tree to use as a control.

10.2.6 The Transfer Rate and Aggregated Transfer Factors into Fruits of ^{137}Cs Applied on Fruiting Mother Shoots Before Sprouting (EXP. 4)

10.2.6.1 Plant Material and Treatments

The experiments were conducted using a 17-year-old "Hiratanenashi" Japanese persimmon tree planted at FTRC. Eight 2-year-old lateral branches, which had several fruiting mother shoots, were selected before sprouting for treatment. In 2020, the apical fruiting mother shoots and their previous year's branches (2-year-old lateral branch) were selected for application of the dis-^{137}Cs solution. In 2021, only apical fruiting mother shoots were selected. The dis-^{137}Cs solution was applied to the front side of the apical fruiting mother shoots and of the 2-year-old branches using a polyamide brush (Fig. 10.4). The [^{137}Cs] was measured in two treated shoots or branches that had been collected a few seconds after treatment, and the RR was calculated. In 2020, 30 days after full blossom the number of fruit set was adjusted to provide a leaf-fruit ratio of 25 leaves per fruit on July 2. No fruit thinning was carried out in 2021.

10.2.6.2 Calculation of the [^{137}Cs] per Unit Area of Contaminated Fruiting Mother Shoot

Before application of the dis-^{137}Cs solution, the length and thickness of the fruiting mother shoot or 2-year-old lateral branch for experiment were measured, and the [^{137}Cs] per unit area of shoots and branches (A_b) was calculated as follows:

$$A_b = AS_{io} \times (0.5 \times \pi \times W \times L)^{-1}, \tag{10.3}$$

where A_b is the amount of ^{137}Cs per unit area of the contaminated shoots and branches (Bq m^{-2}), AS_{io} is the amount of ^{137}Cs in the contaminated shoots and branches (Bq shoot^{-1}), W is the thickness of the contaminated shoots and branches (m), L is the length of the contaminated shoots and branches (m).

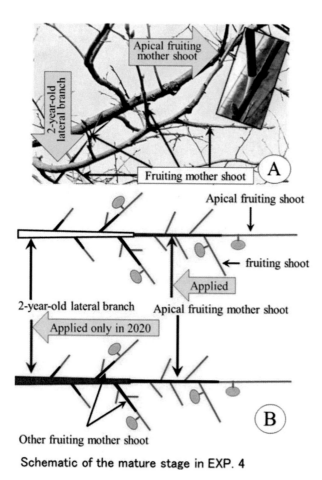

Fig. 10.4 The treatment (**a**) and schematic of the mature stage (**b**) of Japanese persimmon "Hiratanenashi" cultivar in EXP. 4 conducted in 2020 and 2021. Before sprouting in the year of 2020 and 2021, dis-^{137}Cs solution was applied to the front side of the apical fruiting mother shoot and 2-year-old branches (only in 2021) using a polyamide brush (**a**), whereas only apical fruiting mother shoots were applied in 2021

10.2.6.3 Sample Preparation

In 2020, thinned young fruits and mature fruits were collected from the contaminated or uncontaminated apical fruiting mother shoot and the other fruiting mother shoots on 2-year-old branches on July 2 and October 14, respectively. Mature fruits were subdivided into calyx and pulp with skin.

In 2021, on June 25, at the young fruit stage, the contaminated apical fruiting mother shoots and the uncontaminated fruiting mother shoot adjacent to the apical one were collected and divided into wood with bark, shoots, leaves, and young fruits. The contaminated fruiting mother shoots were further subdivided into bark and wood. During the maturity period, the contaminated fruiting mother shoots were collected on October 11 and divided into organs in the same way as in 2020. The replicated samples were combined and then the [^{137}Cs] was measured.

10.2.7 Radiocesium Measurements

After freeze-drying for at least 72 h, all samples from EXP. 1 to EXP. 4 were placed into a 100-mL polypropylene container (U-8 pots) to measure [^{137}Cs]. The [^{137}Cs] in the samples of EXP1, EXP2, EXP4 was measured using a germanium detector (GEM40-76, Seiko EG & G ORTEC, Tokyo, Japan) at Fukushima University. Gamma-ray emissions at 662 keV were measured for 43,200–200,000 s. For the pulse-height analysis, a multichannel analyzer (MCA7, Seiko EG & G ORTEC) was used together with the spectrum analysis software (Gamma Studio, Seiko EG & G ORTEC). The [^{137}Cs] and [^{40}K] in the samples of EXP3 were measured with a germanium detector at the Foundation for Promotion of Material Science and Technology of Japan. Decay correction was not applied as the ^{137}Cs measurement was conducted within 2 weeks of sampling.

10.2.8 Transfer of ^{137}Cs Applied on the Dormant Fruiting Mother Shoot into Organs and Measurement of the Aggregated Transfer Factors of ^{137}Cs into Fruit

The contaminated ^{137}Cs amount (AS, Bq) in each sample due to the application of dis-^{137}Cs solution was calculated using the following equations:

$$AS_i = w_i \times (C_i - C_0) \div 1000, \tag{10.4}$$

where the subscript i represents the type of the organ (e.g., fruit, leaf, calyx, peel, pulp, bark and wood) to be examined, w_i (g DW) is weight of i, C_i (Bq kg^{-1} DW) is the [^{137}Cs] in i, C_0 is the [^{137}Cs] in control of i.

The transfer rate of ^{137}Cs to fruit via contaminated organs (TR_f, %) was calculated using the following equations:

$$TR_f = 100 \times AS_f \div (RR \times AP), \tag{10.5}$$

where AS_f is the amount of ^{137}Cs in fruit (Bq fruit^{-1}), RR is the retained ratio, AP is the amount of applied ^{137}Cs (Bq organ^{-1}).

Based on the equations above, TR_f represents the percentage of ^{137}Cs transported into fruit from the leaves, calyx, or dormant fruiting mother shoots (on which it was deposited).

The aggregated transfer factors of ^{137}Cs applied on the dormant fruiting mother shoot to fruit were calculated using the following equations:

$$T_{ag}{}^{f-b} = (C_f - C_0)A_b{}^{-1}, \qquad\qquad (10.6)$$

where $T_{ag}{}^{f\text{-}b}$ (m^2 kg^{-1}) is the aggregated transfer factors into fruit of ^{137}Cs via the dormant fruiting mother shoot, C_f (Bq kg^{-1} DW) is the [^{137}Cs] in the mature fruit on the contaminated dormant fruiting mother shoot, C_0 is the [^{137}Cs] in control of mature fruit, A_b is the amount of ^{137}Cs per unit area of the contaminated shoots and branches (Bq m^{-2}).

10.2.9   Data Analysis

Analysis of variance (ANOVA) with Tukey pairwise comparison and t-tests were used to test the significance of the differences between comparable [^{137}Cs], TR_f and $T_{ag}{}^{f\text{-}b}$ values. Log-transformed data were used in both the ANOVAs and t-tests.

10.3   Results

10.3.1   Transfer of ^{137}Cs Applied on Calyx and Leaves at the Young Fruit Stage and the Fruit Growing Stage into Dropped Fruit and Mature Fruit of Japanese Persimmon "Hachiya" (EXP. 1)

During the period from the start of July to mid-August, some young unripe fruit had dropped from the tree, so only 2–4 fruits in each treatment remained on the tree up to the harvest time. The TR_f of ^{137}Cs into fruit that had naturally dropped after contamination at the young fruit period was significantly higher due to contamination of the calyx than that for contamination of leaves for both sampling occasions (Table 10.2).

The TR_f of ^{137}Cs derived from EXP. 1 for mature fruit are shown in Table 10.3. The [^{137}Cs] (Bq kg^{-1} DW) measured in the control collected on October 26 were as follows: pulp 5.92, peel 7.20, calyx 10.1 and leaf 10.3 Bq kg^{-1}. The TR_f via calyx was higher after contamination at the young fruit stage, whereas the TR_f via leaves showed the opposite trend. Nevertheless, the TR_f were not significantly different (by ANOVA) between organs, or the stage of contamination. However, ANOVA indicated that there was a possibility of the interaction between contaminated period and organ (0.069 of P value by ANOVA).

Table 10.2 Transfer of ^{137}Cs applied onto calix and leaf during the young fruit stage into Japanese persimmon fruit that had dropped naturally (June drop) at two collecting days

Application days	Applied organ	Replicates (shoot)	Collecting days	Weight of fruit (pulp with peel) (g FW)	Dry matter Pulp with peel (%)	Amount of ^{137}Cs (Bq)[a] Pulp with peel	Calyx	Leaf	TR_f Pulp with peel (%)
24 June–1 July	Calyx	2	4-Jul	7.0 ± 0.3	16.6 ± 0.1	0.29 ± 0.12	3.0 ± 3.0	–	2.47 ± 0.71
	Leaf	3	4-Jul	7.2 ± 2.0	16.8 ± 0.3	0.06 ± 0.04	0.05 ± 0.03	3.7 ± 1.2	0.23 ± 0.17
	Calyx	3	11-Jul	10.3 ± 2.4	15.6 ± 0.7	0.33 ± 0.03	1.4 ± 0.26	–	2.61 ± 0.21
	Leaf	4	11-Jul	9.0 ± 2.5	17.8 ± 4.2	0.24 ± 0.19	0.12 ± 0.03	3.8 ± 1.0	0.96 ± 0.72
P value by	Applied organ								0.003
ANOVA	Collecting days								NS
	Interaction								NS

[a]Arithmetic mean ± standard deviation

Table 10.3 Transfer of ^{137}Cs administered via calyx and leaves to mature Japanese persimmon fruit at two different growth stages

Application days	Applied organ	Replicate (shoot)	Weight of fruit [a](pulp with peel) (g FW)	Dry matter Pulp with peel (%)	Amount of ^{137}Cs (Bq)[b] Pulp with peel	Calyx	Leaf	TR_f Pulp with peel (%)
24 June–1 July	Calyx	3	291 ± 31.5	17.9 ± 0.4	2.0 ± 0.6	0.7 ± 0.3	–	14.9 ± 4.7
	Leaf	2	306 ± 36.3	19.3 ± 0.8	2.5 ± 0.7	0.1 ± 0.0	0.3 ± 0.0	10.2 ± 3.2
1–10 August	Calyx	4	293 ± 25.4	18.3 ± 0.9	2.2 ± 1.0	1.1 ± 0.4	–	10.3 ± 4.6
	Leaf	3	358 ± 73.7	18.9 ± 0.9	2.6 ± 0.8	0.3 ± 0.1	0.5 ± 0.2	16.4 ± 5.0
P value by	Applied organ							NS
ANOVA	Applied days							NS
	Interaction							NS (0.069)

[a]Collected on 26 October
[b]Arithmetic mean ± standard deviation

10.3.2   Effects of the Position Where ^{137}Cs Was Applied onto Leaves on the Transfer into Japanese Persimmon Fruit (EXP. 2)

The result of EXP. 2 is shown in Table 10.4. ^{137}Cs was not detected in the uncontaminated leaves of each treatment. The TR_f of the pulp with peel resulting from contamination of the leaves adjacent to fruit both on the long and on the short fruiting shoot showed the same trend as the [^{137}Cs]. They were significantly higher than those from contamination of the leaves at the middle and top part on a long fruiting shoot (0.01 of P value by ANOVA).

10.3.3   Effects of Fruit Load on the Transfer of ^{137}Cs Applied onto Leaves into Persimmon Fruit (EXP. 3)

The leaf to fruit ratio of "No fruit thinning" was significantly lower than that of the "Fruit thinning" (Table 10.5). The [^{137}Cs] in fruit and the TR_f into fruit of the "Fruit thinning" were higher than that of the "No fruit thinning" (0.02 and 0.055 of P value by t-test, respectively), and the TR_f in short fruiting shoots was an intermediate value. There was no significant difference in fruit weight between both groups. The TR_f of ^{137}Cs into the fruit on the uncontaminated fruiting shoots was less than 2% (Table 10.6). Values of the TR_f of the fruit two nodes below and of the sum of TR_f of the five fruiting shoots in "Fruit thinning" were, respectively, higher than the corresponding TR_f in "No fruit thinning." There was a significant correlation between the amount of ^{137}Cs and ^{40}K as well as of their concentration in the fruit on the contaminated fruiting shoot. The amount of ^{137}Cs as well as ^{40}K in fruits tended to be higher in the "Fruit thinning" (Fig. 10.5). The TR_f of the ^{137}Cs into wood with bark of 2-year-old lateral branch in the "No fruit thinning" group was $1.54 \pm 0.15\%$ (mean ± standard deviation, SD), which was significantly higher than that in the "Fruit thinning" group of $0.30 \pm 0.06\%$. Conversely there was no significant difference in the TR_f into wood with bark of the five fruiting shoots in the "No fruit thinning" and the "Fruit thinning" group, which were $1.18 \pm 0.46\%$ and $1.36 \pm 0.41\%$, respectively.

10.3.4   Transfer of ^{137}Cs Applied on the Dormant Fruiting Mother Shoot into Fruit (EXP. 4)

The leaf to fruit ratio for the 2-year-old branch was 28.6 ± 7.2 where the fruit thinning was applied in 2020 and for the apical fruiting mother shoot was 28.8 ± 8.1 in 2020 and 6.5 ± 3.4 in 2021. A_b was 5.2 ± 4.3 and 4.8 ± 2.3 kBq m^{-2} of branch in

Table 10.4 Concentration, amount, and transfer of ^{137}Cs into Japanese persimmon fruit from leaves contaminated in different positions relative to the fruit

Treatment[c] Position of spiked leaves	Replicate (shoot)	Dry matter Pulp with peel (%)	[^{137}Cs] Pulp with peel[d] (Bq kg^{-1} DW)	Amount of ^{137}Cs (Bq)[f] Pulp with peel	Calyx[e]	Contaminated leaves	TR_f Pulp with peel (%)
Adjacent to fruit on a long shoot	4	16.9 ± 0.5	36.5[a] ± 10.3	1.1[a] ± 0.3	0.06	0.46 ± 0.5	8.4[a] ± 2.6
Middle part in a long shoot	4	17.3 ± 0.5	12.1[b] ± 5.6	0.3[b] ± 0.1	0.02	0.21 ± 0.03	2.7[b] ± 1.2
Top part in a long shoot	4	16.9 ± 0.4	9.6[b] ± 4.1	0.2[b] ± 0.1	0.02	0.29 ± 0.1	2.0[b] ± 0.8
Adjacent to fruit on a short shoot	6	16.9 ± 0.6	41.4[a] ± 24.0	1.1[a] ± 0.5	0.09	0.30 ± 0.3	8.8[a] ± 4.3
P value by ANOVA			0.01	0.004			0.00008

[c] 12.5 Bq of ^{137}Cs was retained after applying 20.3 Bq of ^{137}Cs onto two leaves
[d] Different letters indicate a significant difference at $P \leq 0.05$ by Tukey's test
[e] Measured by combining samples into one sample for the treatment
[f] Arithmetic mean ± standard deviation

Table 10.5 Leaf to fruit ratio on the fruiting mother shoot and [^{137}Cs] of the mature fruit on the contaminated fruiting shoot

Treatment	Replicate (branch)	Fruiting mother shoot			Pulp with peel of fruit on contaminated fruiting shoot			
		Number of fruit	Number of leaf	Leaf to fruit ratio	Number of leaf	Weigh (g FW)	Dry matter (%)	[^{137}Cs] (Bq kg^{-1} DW)
No fruit thinning	1	5	33	6.6	7	238	19.9	22.1
	2	5	37	7.4	9	184	20.2	19.5
	3	5	41	8.2	8	213	20.6	22.2
	4	5	29	5.8	7	236	17.8	19.9
	Mean ± SD[a]		35.0 ± 5.2	7.0 ± 1.0	7.8 ± 1.0	218 ± 25	19.6 ± 1.2	20.9 ± 1.4
Fruit thinning	1	3	47	15.7	5	184	19.4	24.6
	2	3	33	11.0	6	231	18.0	30.3
	3	3	27	9.0	9	252	18.2	43.3
	4	3	35	11.7	7	255	18.0	35.4
	Mean ± SD		35.5 ± 8.4	11.8 ± 2.8	6.8 ± 1.7	230 ± 33	18.4 ± 0.7	33.4 ± 7.9
P value by t-test			NS	0.02	NS	NS	NS	0.02

[a]Arithmetic mean ± standard deviation

Table 10.6 TR_f of ^{137}Cs into fruit for "No fruit thinning" and "Fruit thinning" in contaminated shoots and into fruits borne by branches one or two nodes away from the contaminated central one

Treatment	Replicate (Branch)	TR_f of pulp with peel (%)							Short fruiting shoot
		Long fruiting mother shoot							
		Position of fruiting shoot							
		2 Nodes above	1 Node above	Applied	1 Node below	2 Nodes below	Total	
No fruit thinning	1	0.6	ND	11.8	0.4	0.0	12.8	17.8
	2	0.6	ND	8.0	1.0	0.4	10.0	8.6
	3	ND	ND	11.1	0.3	ND	11.4	11.8
	4	ND	0.1	9.1	2.1	–	11.3	
	Mean ± SD[a]	0.29 ± 0.34	0.08 ± 0.05	10.0 ± 1.8	0.93 ± 0.86	0.15 ± 0.25	11.4 ± 1.2	12.8 ± 4.7
Fruit thinning	1	2.6		10.1		0.6	13.3	
	2	0.8		14.7		1.6	17.1	
	3	0.9		24.1		2.0	27.0	
	4	0.2		19.5		1.9	21.6	
	Mean ± SD	1.13 ± 1.05		17.1 ± 6.1		1.53 ± 0.63	19.7 ± 5.9	
P value by t-test		–		NS (0.055)		0.017	0.017	

[a]Arithmetic mean ± standard deviation

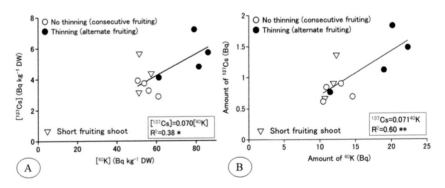

Fig. 10.5 Relationship between [^{137}Cs] and [^{40}K] (**a**), and the amount of ^{137}Cs and ^{40}K (**b**) in fruit on the fruiting shoot contaminated dis-^{137}Cs. * and ** indicate a significant correlation at $P \leq 0.05$ and 0.01 by ANOVA

2020 and 2021, respectively. The TR_f of ^{137}Cs for fruit on the contaminated apical fruiting mother shoot at young and mature stage was $0.64 \pm 0.27\%$ and $1.3 \pm 0.6\%$ in 2020, which was significantly higher than that of ^{137}Cs into fruit on the uncontaminated 2-year-old lateral branch, which was $0.30 \pm 0.27\%$ and $0.29 \pm 0.22\%$ (Fig. 10.6a). In contrast, there was no significant difference in the

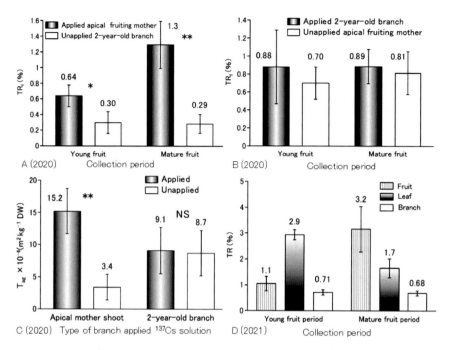

Fig. 10.6 TR_f of ^{137}Cs applied on the apical fruiting mother shoot (**a**) and the 2-year-old lateral branch (**b**), T_{ag} via the fruiting mother shoot and 2-year-old lateral branch contaminated dis-^{137}Cs before sprouting in 2020 (**c**) and TR of ^{137}Cs into leaf, fruit and branch on and of the contaminated apical fruiting mother shoot in 2021 (**d**). The TR into branch in (**d**) was combined both TR of the shoot, of which was bark and wood, and of the wood of the contaminated apical fruiting mother shoot except for bark on which surface remains the contaminated ^{137}Cs. In graph C "Unapplied" at the label of "Apical mother shoot" and "2-year-old branch" represent fruiting mother shoots other than "Apical mother shoot" and an apical fruiting mother shoot. In graph D, "branch" represents the TR except for bark of contaminated branch/shoot. * and ** indicate a significant correlation at $P \leq 0.05$ and 0.01 by ANOVA. Vertical bar represents arithmetic standard deviation

TR_f of the fruit from the contaminated 2-year-old lateral branch between the uncontaminated apical fruiting mother shoot and the other fruiting mother shoot at young and mature period in 2020 (Fig. 10.6b). The T_{ag}^{f-b} of the mature fruit on the contaminated apical fruiting mother shoot was significantly higher than that in fruit on the corresponding uncontaminated fruiting mother shoots, whereas there was no significant difference in the T_{ag}^{f-b} of the mature fruit on the apical fruiting mother shoot and other fruiting mother shoots on the contaminated 2-year-old branch (Fig. 10.6c). In 2021, the TR_f of ^{137}Cs for mature fruit on the contaminated apical fruiting mother shoot was $3.16 \pm 1.75\%$, which was higher than $1.05 \pm 0.56\%$ for young fruits (Fig. 10.6d). The T_{ag}^{f-b} of the mature fruit was $8.0 \pm 4.5 \times 10^{-4}$ m^2 kg^{-1} DW. The TR_f of ^{137}Cs for the mature fruit via the contaminated apical fruiting mother shoot was significantly higher in 2021 than in 2020 (0.025 P value by t-test).

10.4 Discussion

Because RCs was deposited directly onto fruit tree surfaces (Sato et al. 2015b), it was important to evaluate how much RCs was transferred via leaves to fruits after bud burst after the FDNPP accident.

The calyx of Japanese persimmon is larger than that of other fruit trees and plays an important role in fruit enlargement as the respiratory organ of fruit. Calyx lobes fall during the young fruit stage. The shape of the calyx resembles a dish after the fruit growing stage, because the calyx lobes become erect, so RCs in raindrops intercepted by leaves and branches would tend to accumulate in/on the calyx (Fig. 10.7). Previous studies on the transfer of dis-^{137}Cs from cotton contaminated with ^{137}Cs solution onto the calyx of mature Japanese persimmon fruit (cultivar 'Hiratanenashi') reported that TR_f were 2.3% in the "Immature period" during 7–40 days after full blossom and 16.9% in the "Mature period" during 41 days after full blossom to harvest (Sekizawa et al. 2016a, b). The data derived from this study are not consistent with these findings (Table 10.1) because of differences in the method applied and the period during which the contamination occurred. In this study, the contamination period was 23–31 days after full blossom in the young fruit stage and 61–75 days after full blossom in the fruit growing stage. These time periods were shorter than that of Sekizawa et al. and the contamination method involved direct application of ^{137}Cs, whereas Sekizawa et al. used an indirect treatment via cotton. Before 20 days after full blossom most calyx lobes fall, during which dis-^{137}Cs from cotton may have been lost from the calyx. Such losses may explain why TR_f in the "Immature period" on the experiment of Sekizawa et al. was

Fig. 10.7 Stemflow stayed on the calyx of "Hachiya" persimmon

lower than that reported in this study. Anyway, further studies are needed to clarify the transfer of ^{137}Cs into fruit via calyx.

The TR_f of ^{137}Cs from contaminated calyx and leaves in the young fruit stage into fruit that had dropped naturally in EXP. 1 was less than 3% (Table 10.2), whereas the TR_f into mature fruit was three fold higher at >10% (Tables 10.1 and 10.3). In EXP. 4. the TR_f of ^{137}Cs from contaminated apical fruiting mother shoots into fruit before sprouting increased from the young fruit to the mature fruit, whereas for leaves, it decreased during the same period (Fig. 10.6d). Therefore, the data indicates that translocation of ^{137}Cs to fruit via the leaves and bark consistently increases after the fruit growing stage. The TR_f of ^{137}Cs from contaminated leaves in the young fruit stage in Japanese persimmon was lower than that reported for fruit after the fruit growing stage (Table 10.1).

These results are corroborated by studies of Pröhl et al. (2003) who found that the transfer rate to fruits via leaves in apples was higher after spraying a ^{137}Cs solution during the fruit growing stage than at the young fruit stage. Other previous data also shows that for grape, the translocation of ^{137}Cs absorbed from the aboveground part of the tree to fruits increased after the fruit growing period (Madoz-Escande et al. 2002). Furthermore, Davies et al. (2006) showed that two potassium transporters in grape are expressed most highly in the berry skin during the pre-véraison period (a stage at which berries begin to soften and change color). Overall, these studies indicate that the TR_f of ^{137}Cs in fruit due to deposition onto leaves is affected by the time at which contamination occurs, and that the extent of accumulation (sink strength) of ^{137}Cs increases after the fruit growing stage. These results are consistent with other relevant data, which showed that translocation of photosynthetic products from leaves to fruits accelerated after the fruit growing stage (Pavel and DeJong 1995; Yamamoto 2001; Zhang et al. 2005; Beshir et al. 2017).

There was a significant correlation between ^{40}K and ^{137}Cs in the mature fruit in EXP. 3. Moreover, the correlation was higher for the total Bq amount than for activity concentration. Amount represents sink strength and activity concentration represents sink activity (Fig. 10.5). K, an analogue of Cs, promotes the translocation of photosynthetic products from leaves to fruit (Eplon et al. 2015), which explains why the TR_f of ^{137}Cs in fruit via leaf also increased after the fruit growing period.

The effect of the distance between fruit and the leaves as the source of ^{137}Cs was shown in the experiment with Japanese persimmon "Hiratanenashi," where the TR_f after contamination of the leaves adjacent to the fruit was much higher than that after contamination of leaves located further away at the top of the long shoot (Table 10.4). These observations indicate that the relationship between supply and demand of photosynthetic products on the fruiting shoot influenced the sink and source relationship of ^{137}Cs between leaf and fruit. Katana et al. (1988) placed a ^{134}Cs solution onto leaves at the top of the apple branch. Consequently, 19–42% of ^{134}Cs administered was transferred to fruit on the same contaminated branch, whereas a small amount of ^{134}Cs (at 0.01–0.4%) was detected in fruit at the base of the 25 cm length lower branch. They concluded that the relationship between the contaminated leaves and fruit had a tendency that the transport of radiocesium deposited on the leaves to the fruits is restricted within the branches on which contaminated leaves are growing, and it is difficult to migrate into the fruits on the different branches. Carini et al. (1999) compared the transfer of ^{134}Cs from contaminated leaves to fruits between contaminated branches and uncontaminated branches in grapes, apples, and pears. The transfer factor of ^{134}Cs in fruit of the contaminated branch, which was defined by fruit concentration (Bq kg^{-1} DW) divided by Bq administered per plant, was 1.33, 5.11 and 0.69; this was significantly higher than that of the uncontaminated branch of 0.098, 0.1 and not detected for grapes, apples, and pears, respectively. These data showed that there was a close relationship between source and sink for ^{134}Cs on the contaminated branch.

At the time of the FDNPP accident, the Japanese persimmon was in the dormant stage prior to bud burst and ^{137}Cs on the contaminated bark of the aboveground part of the fruit tree migrated directly into the tree. After bud burst, Sato et al. (2021) reported that leaves were also contaminated with ^{137}Cs via raindrops in the canopy during or after rainfall. K^{+} is readily absorbed by leaves (Franke 1967; Schönherr and Luber 2001) and Mcfarlane (1974) reported that Cs^{+} penetrates through the cuticle more easily than K^{+}. K^{+} penetrates through the fine pores in the cuticle on the leaf surface, thereby reaching parenchyma cells and is then transported via the vascular system. K plays a key role in elevating the translocation of assimilates from leaves to fruit after the fruit growing period (Eplon et al. 2015). Conversely, K leaches through the cuticle easily from leaves due to rainfall (Tukey and Mecklenburg 1964; Mecklenburg and Tukey 1964; Potter et al. 1991), which probably also occurs for radiocesium.

In this study, the effect of the fruit load on TR_f differed depending on which organ was contaminated. In EXP. 3, ^{137}Cs administration onto leaves of fruiting shoots that had been subjected to fruit thinning just before the contamination treatment on July 31 resulted in a higher value of the TR_f than when thinning was not applied. Similarly, a higher amount of ^{40}K was translocated into fruits on the contaminated long fruiting shoot in the "Fruit thinning" scenario (showing that the sink activity and strength of these fruits works effectively) compared with the "No fruit thinning" shoot. The difference indicates that the accumulation by these fruits was suppressed possibly by competition with other fruits on the uncontaminated shoots that had developed on the same mother shoot.

In EXP. 4, the administration of ^{137}Cs to a dormant apical fruiting mother shoot without fruit thinning resulted in a higher TR_f than that in the year with fruit thinning. The number of fruits per shoot in the experiment in 2021 was approximately four times higher than that of the experiment in 2020. The difference in fruit load between the experimental years in EXP. 4 is larger than that between the treatments in EXP. 3. This suggests that the higher TR_f in the experiment without fruit thinning was due to the different mass of the fruits. Conversely, there was a significantly higher [^{137}Cs] in EXP. 3 (Table 10.5) and $T_{ag}{}^{f-b}$ in EXP. 4 (Fig. 10.6c, d) in the experiment with fruit thinning. These initial data indicate that refraining from heavy fruit thinning may be a useful remediation option that could reduce the [^{137}Cs] in Japanese persimmon fruits.

In EXP. 4 conducted using the same leaf to fruit ratio, the transfer of contaminated ^{137}Cs was compared between fruits on the apical fruiting mother shoot and the other fruiting mother shoots developed from the same 2-year-old lateral branch. When dis-^{137}Cs solution was applied to the apical fruiting mother shoot, the TR_f of the combined mature fruits collected from the other fruiting mother shoots was less than a quarter of that for the fruit on the contaminated apical mother shoot (Fig. 10.6a). In contrast, when dis-^{137}Cs solution was applied to the 2-year-old lateral branches, the TR_f and the T_{ag}^{f-b} into fruit were similar between the fruit on the apical fruiting mother shoot and the other fruiting mother shoots (Fig. 10.6b, c).

These results indicate that the transfer of ^{137}Cs to fruits grown on fruiting mother shoots developed from the same contaminated 2-year-old lateral branch is similar, whereas the transfer of ^{137}Cs from a contaminated apical fruiting mother shoot to fruits grown on the other shoots developed from the same 2-year-old lateral branch is limited. Furthermore, [^{137}Cs] of fruit on the contaminated fruiting shoot in the fruit thinning was higher than that with no fruit thinning in EXP. 3 (Table 10.6), and T_{ag}^{f-b} in 2020 when the fruit was thinned was higher than that in 2021 where the fruit was not thinned in EXP. 4. These data are consistent with the experimental results assuming that the transfer of ^{137}Cs is controlled by the sink activity of fruits.

T_{ag}^{f-b} values derived in EXP. 4 were of the order of magnitude of the 10^{-4} m^2 kg^{-1} DW. Unfortunately, there are no previous data reported for Japanese persimmon after the Chernobyl accident, and only a few data for T_{ag} in fruit after the FDNPP accident. Renaud and Gonze (2014) reported T_{ag} (m^2 kg^{-1} FW), which vary in a range of 10^{-4} to 10^{-3} m^2 kg^{-1} FW. For persimmon T_{ag} values of 9.7×10^{-4} m^2 kg^{-1} FW was previously reported by Tagami and Uchida (2014). The value of 15.2×10^{-4} m^2 kg^{-1} DW of T_{ag}^{f-b} in EXP. 4 is the equivalent of 2.4×10^{-4} m^2 kg^{-1} FW of T_{ag}^{f-b}. Although the leaf area of the contaminated leaves was not measured in EXP. 3, the T_{ag} into fruit via leaves calculated assuming a leaf area of 72 cm^2 (Fujimoto and Tomita 2000) was 10^{-3} m^2 kg^{-1} FW. Therefore, T_{ag} into fruit via dormant branches can be tentatively estimated to be one order of magnitude lower than T_{ag} into fruit via leaves.

This study has shown that T_{ag}^{f-b} for the fruit on the apical mother shoot were approximately twice as high as for fruits on other fruiting mother shoots developed on the same 2-year-old branch contaminated during dormancy. However, similar experiments on fruiting mother shoots on 2-year-old branches other than apical fruiting mother shoots have not yet been conducted. The currently available data suggest that fruit load is likely to be one of the main factors affecting variation in the extent of [^{137}Cs] transfer into fruits within the same lateral branch.

To our knowledge, this was the first reported study that has quantified the transport of ^{137}Cs into fruit via the dormant lateral branch. Further studies are needed to clarify the dynamics of ^{137}Cs inward migration into deciduous fruit trees that have been externally contaminated during dormancy and other growth phases.

10.5   Limitations

The dis-^{137}Cs solution used in EXP. 4 contained up to approximately 0.06 Bq mL^{-1} of ^{40}K (0.2% as natural K), but no ^{40}K was detected in the dis-^{137}Cs solution used epiphytic moss growing on rock. Natural K is present on leaves and bark surfaces by leaching, and inward-migration of RCs from the tissue surface is a physical permeation phenomenon, so inward-migration of dis-^{137}Cs likely is not affected by K in the applied solution unlike absorption via the roots. No significant differences in ^{40}K activity concentrations in organs between of the control group and the experimental group were found on each sampling date, whereas a considerable amount of applied ^{137}Cs was detected in each experiment of this study. While K in the applied dis-^{137}Cs seems to have negligible relevance to the result in each experiment, utilizing the parameter calculated in EXP. 4 needs to take into account the precondition which is the experiment using the dis-^{137}Cs solution contained up to 0.2% as natural K.

Acknowledgements   The author wishes sincerely to thank Prof. Brenda Howard and Prof. Franca Carini for their useful comments and for improving the English in the manuscript. Part of the research results was obtained from projects conducted at the FTRC.

References

Aoki D, Asai R, Tomioka R, Matsushita Y, Asakura H, Tabuchi M, Fukushima K (2017) Translocation of ^{133}Cs administered to *Cryptomeria japonica* wood. Sci Total Environ 584: 88–95

Beshir WF, Mbong VBM, Hertog MLATM, Geeraerd AH, Ende WV, Nicolaï BM (2017) Dynamic labeling reveals temporal changes in carbon re-allocation within the central metabolism of developing apple fruit. Front Plant Sci 8:1785. https://doi.org/10.3389/fpls.2017.01785

Carini F, Lombi E (1997) Foliar and soil uptake of ^{134}Cs and ^{85}Sr by grape vines. Sci Total Environ 207:157–164

Carini F, Scotti IA, Montruccoli M, Silva S (1996) ^{134}Cs foliar contamination of vine: translocation to grapes and transfer to wine. In: Gerzabek M (ed) International symposium on radioecology 1996: 10 years terrestrial radioecological research following the Chernobyl accident. Austrian Soil Science Society, Vienna, pp 163–169

Carini F, Scotti IA, D'Alessandro PG (1999) ^{134}Cs and ^{85}Sr in fruit plants following wet aerial deposition. Health Phys 77:520–529

Chauhan N (2014) Pollination studies in relation to fruit drop in persimmon (*Diospyros kaki* Thunb.) cv. Hachiya. Thesis. Dr. Y. S. Parmar University of Horticulture and Forestry, Solan

Davies C, Shin R, Liu W, Thomas MR, Shachitman DP (2006) Transporters expressed during grape berry (*Vitis vinifera* L.) development are associated with an increase in berry size and berry potassium accumulation. J Exp Bot 57:3209–3216

Eplon E, Cabral OMR, Laclau J, Dannoura M, Packer AP, Plain C, Battie-Laclau P, Moreira MZ, Trivelin PCO, Bouillet J, Gérant D, Nouvellon Y (2015) In situ ^{13}CO$_2$ pulse labelling of field-grown eucalypt trees revealed the effects of potassium nutrition and throughfall exclusion on phloem transport of photosynthetic carbon. Tree Physiol 36:6–21

Faust M (1989) Physiology of temperate zones fruit trees. Wiley, New York

Franke W (1967) Mechanisms of foliar penetration of solutions. Annu Rev Plant Physiol 18:281–300

Fujimoto K, Tomita E (2000) Effect of fruit thinning on fruit quality and yield of Japanese 'Hiratanenashi' persimmon. Bull For Fish 1:55–66

Horii S, Kusaba S, Sekizawa H, Hachinoue M, Hamamatsu S, Matsunami H, Nonaka A, Murakami T (2019) Effect of number of leaves per fruit on radiocesium concentration of Japanese persimmon fruits in maturation period. Bull NARO Agric Res Tohoku Reg 121:39–47

IAEA (International Atomic Energy Agency) (2020) Environmental transfer of radionuclides in Japan following the accident at the Fukushima Daiichi Nuclear Power Plant. IAEA-TECDOC-1927. IAEA, Vienna, pp 68–90

Katana H, Bunnenberg C, Kuhn W (1988) Studies on the translocation of ^{134}Cs from leaves to fruit of apple trees. In: Impact des accident d'Origine nuclearie Sur l'Environment, proceedings of the IV symposium international de Radioecologie de Cadarache, vol 2. Commissariat a l'Energie Atomique, Paris, pp 72–78

Kitajima A, Fujiwara T, Kukizaki T, Ishida M, Soejima Z (1987) Relationship between early fruit drop and dry matter accumulation on bearing shoot in Japanese persimmon (*Diospyros kaki* Thunb.). Sci Rep Kyoto Prefect Univ Agric 39:1–11

Kitajima A, Matsumoto S, Ishida M, Soejima Z (1990) Relationship between dry matter production of bearing shoots and physiological fruit drop of Japanese persimmon by shading treatment. J Jpn Soc Hort Sci 59:75–81

Kuroda K, Kagawa A, Tonosaki M (2013) Radiocesium concentrations in the bark, sapwood and heartwood of three tree species collected at Fukushima forests half a year after the Fukushima Dai-ichi nuclear accident. J Environ Radioact 122:37–42

Lang A (1983) Turgor-related translocation. Plant Cell Environ 6:683–689

Madoz-Escande C, Colle C, Adam C (2002) Evolution of cesium and strontium contamination deposited on vines. Radioprotection 37(1):515–520

Mcfarlane JC (1974) Cation penetration through isolated leaf cuticles. Plant Physiol 53:723–727

Mecklenburg RA, Tukey HB (1964) Influence of foliar leaching on root uptake and translocation of calcium-45 to the stems and foliage of *Phaseolus vulgaris*. Plant Physiol 39:533–536

Ministry of Agriculture, Forestry and Fisheries (2010) Survey on the production of specialty fruit trees. https://www.maff.go.jp/j/tokei/kouhyou/tokusan_kazyu/p001-22-047.xls

Ministry of Agriculture, Forestry and Fisheries (2022) The 88th statistical yearbook of ministry of agriculture, forestry and fisheries (2019–2020). http://www.maff.go.jp/e/data/stat/95the/attach/xls/index156xls

Pavel EW, DeJong TM (1995) Seasonal patterns of nonstructural carbohydrates of apple (*Malus pumila* mill.) fruits: relationship with relative growth rates and contribution to solute potential. J Hort Sci 70:127–134

Possner D, Kliewer W (1985) The localization of acids, sugars, potassium and calcium in developing grape berries. Vitis 24:229–240

Potter CS, Ragsdale HL, Swank WT (1991) Atmospheric deposition and foliar leaching in a regenerating southern Appalachian Forest canopy. J Ecol 79:97–115

Pröhl G, Fiedler I, Koch-Steindl H, Leser C, Treutter D (2003) Interzeption und translokation von radiocäsium bei Obst und Beeren. Bundesministerium für Umwelt. Naturschutz und Reaktorsicherheit, Bonn, pp 6–21

Renaud P, Gonze MA (2014) Lessons from the Fukushima and Chernobyl accidents concerning the ^{137}Cs contamination of orchard fresh fruits. Radioprotection 49(3):169–175

Sato M, Takata D, Abe K, Kawatsu K, Tanoi K, Ohno T, Ohtsuki T, Takase T, Muramatsu Y (2014) Investigation of the radiocaesium migration pathway into the deciduous fruit tree contaminated at the dormant period. Bulletin Fukushima Agriculture Technology Central Radioactive Substances Special Issue: 70–73

Sato M, Abe K, Kikunaga H, Takata D, Tanoi K, Ohtsuki T, Muramatsu Y (2015a) Decontamination effects of bark washing with a high-pressure washer on peach [*Prunus persica* (L.) Batsch] and Japanese persimmon (*Diospyros kaki* thumb.) contaminated with radiocaesium during dormancy. Hort J 84:295–304

Sato M, Takata D, Tanoi K, Ohtsuki T, Muramatsu Y (2015b) Radiocaesium transfer into the fruit of deciduous fruit trees contaminated during dormancy. Soil Sci Plant Nutr 61:156–164

Sato M, Takase T, Yamaguchi K (2017a) Development of methods for collecting the stemflow on the trunk of trees contaminated with radioactive fallout. J Agric Meteorol 72:82–87

Sato M, Takase T, Yamaguchi K (2017b) Effects of bark washing and epiphytic moss on ^{137}Cs activity concentration in bark and stemflow in Japanese persimmon (Diospyros kaki Thunb.). J Environ Radioact 178:360–366

Sato M, Matsuoka K, Takase T, Kobayashi IN, Kikunaga H, Takata D, Tanoi K, Ohtsuki T, Kusaba S, Yamaguchi K (2019a) Vertical migration of Cs-137 in Japanese orchards after the Fukushima Daiichi nuclear power plant accident. Hort J 88(2):150–163

Sato M, Kuwana A, Minami H, Watanabe Y, Takata D, Ohno T, Kikunaga H, Ohtsuki T (2019b) Long-term changes of ^{137}Cs activity concentration in fruit and leaf of deciduous fruits during both period of young and mature fruit stage, proceedings of the 20th workshop on environmental radioactivity. High Energy Accelerator Research Organization, Tsukuba, pp 143–148

Sato M, Takata D, Watanabe Y, Takase T, Yamaguchi K (2021) Evaluation of the deposition of ^{137}Cs in Japanese persimmon trees and yuzu trees from rainfall by collecting raindrops with sphagnum pads. Act Hort 1312:581–589

Schönherr J, Luber M (2001) Cuticular penetration of potassium salts: effects of humidity, anions, and temperature. Plant Soil 236:117–122

Sekizawa H, Sato M, Aihara T, Murakami T, Hachinohe M, Hamamatsu S (2016a) Translocation of the radioactive cesium via the calyx in persimmon fruit. Radioisotopes 65:129–135

Sekizawa H, Sato M, Aihara T, Murakami T, Hachinohe M, Hamamatsu S (2016b) Translocation of the radioactive caesium via the calyx in persimmon fruit (part 2) addition amount of radioactive caesium to the calyx and transfer rate to pulp. Radioisotopes 65:507–515

Sobashima Y, Takagi M (1968) Investigation on the cause and control of physiological dropping in the Japanese persimmon fruits: the time of fruits dropping and abscission layer formation. Sci Repo Kyoto Prefect Univ Agric 20:1–11

Tagami K, Uchida S (2014) Concentration change of radiocaesium in persimmon leaves and fruits. Observation results in 2011 Spring–2013 Summer. Radioisotopes 63(2):87–92

Tagami K, Uchida S, Ishii N, Kagiya S (2012) Translocation of radiocesium from stems and leaves of plants and the effect on radiocesium concentrations in newly emerged plant tissues. J Environ Radioact 111:65–69

Takata D (2013) Distribution of radiocaesium from the radioactive fallout in fruit trees. In: Nakanishi TM, Tanoi K (eds) Agricultural implications of the Fukushima nuclear accident. Springer, Tokyo, pp 143–162

Tukey HB, Mecklenburg RA (1964) Leaching of metabolites from foliage and subsequent reabsorption and redistribution of the leachate in plants. Am J Bot 51:737–742

Van Goor BJ, Van Lune P (1980) Redistribution of potassium, boron, iron, magnesium and calcium in apple trees determined by an indirect method. Physiol Plant 48:21–26

Warren WJ (1972) In crop processes in controlled environment. Academic Press, New York, pp 7–30

Watanabe T, Hoshi M (1981) Studies of growth and fruiting in 'Hiratanenashi' (Japanese persimmon) (2): histological observation on the calyx of persimmon fruit. Bull Yamagata Univ Agric Sci 8:665–681

Westwood MN (1993) Temperate-zone pomology. Physiology and culture. Timber Press, Inc., Portland, pp 185–198

Yamamoto T (2001) Translocation of ^{13}C-photosynthates among 2-year-old branches during the rapid growth stage of cherry, pear, apple and persimmon fruit. J Jpn Soc Hort Sci 70:170–177

Zehnder HJ, Kopp P, Eikenberg J, Feller U, Oertli JJ (1995) Uptake and transport of radioactive cesium and strontium into grapevines after leaf contamination. Radiat Phys Chem 46(1):61–69

Zhang C, Tanabe K, Tamura F, Itai A, Wang S (2005) Partitioning of C-photosynthate from spur leaves during fruit growth of three Japanese pear (Pyrus pyrifolia) cultivars differing in maturation date. Ann Bot 95:685–693

Chapter 11
Progress Toward Managing Radiocesium Contamination in Orchards

Daisuke Takata

11.1 Introduction

Prior to the major earthquake in Fukushima Prefecture in 2011, the prefecture had 7177 ha of land under fruit tree cultivation and was the prefecture with the largest area used for fruit trees in Japan (Ministry of Agriculture, Forestry and Fisheries (Japan) 2018). Fukushima Prefecture was ranked second in Japan with respect to the hectarage used for peach cultivation. According to the Special Fruit Production Survey, 2010 (Ministry of Agriculture, Forestry and Fisheries (Japan) 2012), it ranked first among Japanese prefectures for the largest production of dried persimmons, much of which was "anpo-gaki," a product of especially high value. The sales of anpo-gaki in 2010 by Date Mirai, a former branch of the Japan Agricultural Cooperatives centered around dates, reached ¥1.9 billion (Komatsu 2014a). Although the retail prices of most types of fruit are gradually returning to pre-earthquake levels (Komatsu 2014a, b; Takata 2016), they have still not completely recovered. Nevertheless, Fukushima Prefecture's position as one of the most important fruit-producing prefectures in Japan has not changed since the Tokyo Electric Power Co.'s Fukushima Daiichi Nuclear Power Plant (FDNPP) 2011 disaster.

The effects of radiocesium from the FDNPP disaster on agricultural products as reference materials for the time of the disaster have been estimated. Numerous studies have reported on the effects of radiocesium on annual crops. These studies include (1) surveys of the effects of radioactive fallout from atmospheric nuclear tests; (2) dynamic surveys after the Chernobyl disaster; and (3) studies in experimental environments such as application-using reagents. Although the transfer

D. Takata (✉)
Faculty of Food and Agricultural Sciences, Fukushima University, Fukushima City, Fukushima, Japan
e-mail: r841@ipc.fukushima-u.ac.jp

© The Author(s) 2023
T. M. Nakanishi, K. Tanoi (eds.), *Agricultural Implications of Fukushima Nuclear Accident (IV)*, https://doi.org/10.1007/978-981-19-9361-9_11

coefficient of radiocesium concentration in the soil and perennial fruit crops has been studied (IAEA 2003, 2010), overall, fewer radiocesium-related studies have been conducted on perennial crops than on other crops. One major difference between the FDNPP and Chernobyl disasters was the season in which they occurred; the FDNPP disaster occurred while deciduous fruit trees had no leaves, suggesting that radiocesium fallout would have landed on the tree bodies (i.e., trunks and main branches). With respect to the transfer of radiocesium to inside the fruit-tree bodies, the effects of direct contamination of the trunk and transfer of radiocesium from the soil must be investigated. As fruit trees are long-lived, they have many similarities with timber trees in terms of life cycle and growth. More studies on radiocesium have been reported in forestry than in fruit trees; therefore, forestry studies had to be referred to when clarifying the dynamics of radiocesium in fruit trees. However, in timber plantations, the surface layers of the soil form the litter layer, which contains large quantities of fallen leaves and other organic matter, and this layer plays important roles in root growth and nutrient uptake and cycling. Grass grows on the soil surface in orchards, and, indeed, leaves that fall from trees are often removed by the workers to prevent tree diseases. Considering that orchards are maintained with far more human effort than timber plantations, fruit trees may exhibit vastly different radiocesium dynamics.

In summary, numerous issues related to radiocesium dynamics in orchards were not elucidated before the FDNPP disaster, and there may be numerous differences between perennial crops and timber trees. The reference information available offers starting points to elucidate radiocesium dynamics in fruit trees and develop technologies relevant to this issue.

11.2 FDNPP Disaster and Radioactive Contamination of Orchards

11.2.1 Radiocesium Contamination of Fruit and Leaves in the First Year After the FDNPP Disaster

High concentrations of radiocesium, exceeding the predictions based on coefficients of transfer from the soil at the immature fruit stage (IAEA 2003, 2010), were found in deciduous fruit trees in Fukushima Prefecture after the FDNPP disaster. Radiocesium accumulates readily in immature stages of fruit such as ume (*Prunus mume*: a small, plum-like fruit, in which the flowers were directly contaminated), and chestnuts (which has edible seeds) (Hamada et al. 2012). However, the FDNPP disaster occurred before budbreak, and radiocesium was detected at above the detection limit in the usual deciduous-tree fruits in Fukushima Prefecture that do not belong to the above groups (Suzuki et al. 2018). Radiocesium was detected in all five of the most common fruit species in Fukushima Prefecture in 2011, that is, peaches, grapes, Japanese pears, apples, and persimmons (Abe et al. 2014; Yuda et al. 2014).

With respect to the effects of tree age on radiocesium concentration in fruit, a study was performed in which peach trees of the "akatsuki" variety, 12- and 18-year old, were compared with peach trees of the "yuzora" variety, 6- and 18-year old, and the radiocesium concentration in mature fruit was found to be substantially higher in the older trees (Abe et al. 2014). In addition, differences were found between radiocesium concentrations in the following seven species of fruit: ume, cherry, peach, grape, pear, apple, and persimmon. The radiocesium concentrations in mature fruit differed according to tree age and species, suggesting that above-ground tree morphological factors, such as the spread of the crown, may have affected fruit radiocesium concentration in the first year after contamination.

11.2.2 Radiocesium Contamination of the Bark and Tree in the First Year After the Disaster

Radiocesium concentrations were compared in (1) soil below peach trees planted in pots, with the soil surface covered before the FDNPP disaster; (2) soil below uncovered trees; and (3) the fruit (Takata et al. 2012a). The radiocesium concentration in covered soil decreased by one-sixth, but, with or without covering, the radiocesium concentration in the roots was below the limit of detection. The radiocesium concentrations in fruit with and without covering were 26.6 and 27.8 Bq kg^{-1} fresh weight, respectively, the difference between these values being within the range of measurement error. In addition, the radiocesium content was mostly above ground, and there was no difference with and without covering.

In peach trees, the transfer of radiocesium from the above-ground part of the tree to inside the tree has been investigated (Takata et al. 2012b; Takata 2013; duplicated figure: Fig. 11.1). Samples of the trunks were collected in January 2012, and images of the species of radioactive nuclei were prepared using an imaging plate. Figure 11.1 shows a superimposed photograph and the original image, and the black spots are the loci where radioactive nuclei were detected. Photographs were only taken of the outermost layer of the peach tree bark, and no photographs were taken for loci immediately below that. Radiocesium concentrations were measured using germanium semiconductor detectors in the epidermis and directly under the epidermis, and it was found that extremely high concentrations of radiocesium were present in the epidermis. Therefore, the radiocesium deposited on the tree may be heterogeneously distributed in the bark, with a very high concentration in the thin, outermost layer of the bark.

11.2.3 Radiocesium Contamination of Soil and Grass in the First Year After the Disaster

In orchards in Fukushima Prefecture, grass is grown using the usual ground management methods. In orchards, unlike other cultivated land, branches are at least 3 m

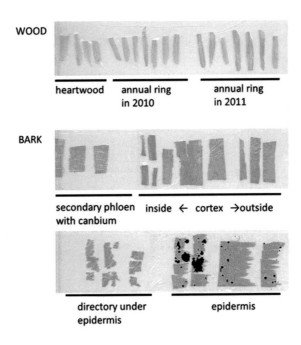

Fig. 11.1 Imaging plate of trunk in 'Akatsuki' peach. From Takata (2013)

high. Consequently, before radioactive fallout reaches the ground in an orchard, it may be affected by stagnant air, owing to the trees, and radiocesium capture, which may result in differences in the amount of radiocesium deposited on the ground surface. In this context, an investigation of the distribution of mean radiocesium concentrations in the top 5 cm of soil (Sato et al. 2015) revealed that the radiocesium concentration in the 5-cm layer in orchards showed two- to fivefold differences between tree crowns. This difference was greater in higher-density apple orchards than in peach orchards, indicating that horizontal differences in the distribution of radiocesium concentration in soil are affected by planting density.

11.3 Elucidation of Radiocesium Uptake by Tree Bodies and Fruit, and Transfer Routes

11.3.1 Radiocesium Transfer from the Bark to Inside

The route of penetration of the tree through the bark by radiocesium has been hypothesized to involve some mode of transfer from the lenticels on the bark to the phloem and/or xylem (Takata 2013). Lenticels on peach trees are often split, and radiocesium can penetrate the body from these sites. Physical splitting on branches not only occurs at lenticels, and intense images of contamination were also obtained at sites with pruning scars, that is, cavities. Physical irregularities are marked at such sites, and dust containing radiocesium tends to accumulate, such that lenticel tissues are likely to be one of the penetration routes for radiocesium into the trees. The lenticels on the outer bark are continuous with the ray tissue in the inner bark. The ray tissue is composed of parenchyma cells that grow in a radiating pattern and pass through the phloem, cambium, and wood. In addition to having an aeration function for the sapwood (Mio and Matsumoto 1979), in recent years, the ray tissue has been acknowledged to have nutrient storage functions (Islam and Begum 2012). Studies on nutrient and water exchange functions with the wood and phloem are in progress to elucidate these mechanisms (Fromm 2010; Pfautsch et al. 2015; Van Bell 1990). In studies with sugi (*Cryptomeria japonica*) and konara (*Quercus serrata*), ^{133}Cs applied to the bark was detected in the wood (Mahara et al. 2017). Visualization assays revealed that ^{133}Cs localized in the ray tissue of sugi, and ^{133}Cs in the bark was transported into the wood via the ray tissue (Aoki et al. 2017). In the FDNPP disaster, it is probable that radiocesium passed through the bark of deciduous trees during the dormant period and was then transferred directly into the body of the fruit trees.

11.3.2 Distribution of Radiocesium in Peach Tree Bodies Five Months After the Disaster

A series of surveys was performed on peach trees to study the distribution of radiocesium in fruit-tree bodies in the first year after the disaster (Takata et al. 2012a, b, c, d). In akatsuki peach trees in orchards that received fallout, at least 20% of the radiocesium, as becquerel equivalents, was in the leaves and fruit (Fig. 11.2). Leaves that fell in November had lower radiocesium concentrations than those that fell in August (Takata 2013), and fresh growth after harvest showed secondary elongation, such that 15–20% of the radiocesium was eliminated from the tree. It has been reported that, in the standard pruning methods for peach trees, 56.6% of the length is removed in the winter pruning season (Takata et al. 2008), but in the first year after the disaster, approximately 30% of the radiocesium in the branches (becquerel equivalents) was eliminated from the tree. By calculating the amount of material collected from the ground (fallen leaves), the amount pruned, and the increase in wood thickness, the amount eliminated from the tree each year can be estimated. The bark of the trunk, which is vertical, has a lower radiocesium concentration than the ground and the main branches, which grow approximately horizontally (Takata et al. 2012b, d). Therefore, when such branches were removed in the winter pruning season, the vigorous renewal of the lateral branches from adventitious buds had a positive effect on tree decontamination. High-pressure washing of the trees to artificially remove radiocesium from the tree bodies in Fukushima Prefecture, from December 2011 to February 2012, considerably reduced the radiocesium concentration.

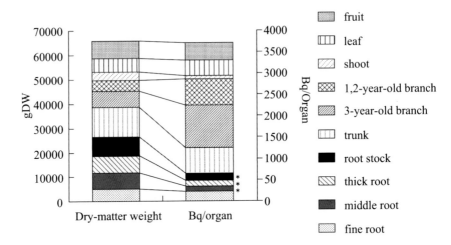

Fig. 11.2 Dry matter weight and Cs content for every organ in 'Akatsuki' peach tree. From Takata (2013). *Under detection limit. Value means detection limit × dry weight

In a study performed with the above group the year after the disaster, in which contaminated peach trees were transplanted in uncontaminated soil (Takata et al. 2014; Takata 2016), only approximately 2% of the radiocesium in the tree was transferred to fruit and other newly grown organs, clearly showing there were major differences in radiocesium transfer and distribution between the year of the disaster and subsequent years. This shows that the efficacy of decontamination of the tree is greatest during the first year after the disaster. It is therefore imperative to perform decontamination, such as washing of trees, promptly after a disaster.

Sato et al. (2014) performed a study in which 6-year-old persimmon trees, grown before the FDNPP disaster, were replanted in pots filled with radiocesium-contaminated soil. Of the ^{137}Cs absorbed from the soil after 3 years, 38.9 ± 6.9% (mean ± standard deviation) was above ground, and 61.1 ± 6.9% was below ground, which was approximately the opposite of the distribution of 70.0% above ground and 30.0% below ground for trees cultivated outdoors in the first year after the disaster. In a damage survey performed for 3 years after the disaster in yuzora peach trees planted before the disaster in experimental plots at the National Institute of Fruit Tree Science, distribution to underground parts was <20% (Takata 2019). In a study with 2-year-old Pinot Blanc grapevines in pots, in which contamination was restricted to the soil or leaves (Carini and Lombi 1997), 42% of uptake from the soil only was distributed to underground parts, whereas proportional distribution to underground parts was 10% when uptake was from the leaves only, clearly indicating that the distribution of radiocesium above and below ground reflects the site of origin of the radiocesium. In the case of fruit trees contaminated by the FDNPP disaster, it is highly probable that radiocesium uptake was directly into the tree above ground, and the amount of radiocesium storage in the tree is therefore greater above ground.

11.3.3 Transfer from Soil to Fruit

Comparison of the bodies of peach trees, in which the soil surface was either covered and uncovered at the time of radiocesium fallout, showed that the radiocesium concentration in soil was reduced by one-sixth by covering and the radiocesium concentrations of the roots and the above-ground parts were similar, whether covered or not (Takata et al. 2012a). When peach trees that received radiocesium fallout were replanted, there were no differences in the radiocesium concentration in the fruit, irrespective of whether the soil used was contaminated or uncontaminated with radiocesium. Horii et al. reported that the transfer coefficients over several years after planting uncontaminated unshu mandarin orange (*Citrus unshiu*), chestnut, and persimmon trees were in the range of 10^{-4} to 10^{-3} (2018). On the basis of a previous report (IAEA 2003), and the present results, it is considered that in fruit trees, the contamination of the soil surface layer has little effect on uptake by the roots.

There have been several reports on the transfer of radiocesium from the soil, including one on fig and grape trees, in which the soil concentration was changed locally (Takata et al. 2013b); a second, with blueberry bushes, on the relationship between the exchangeable potassium concentration and radiocesium concentration in fruit (Iwabuchi 2014); and a third on the cation concentration in the soil solution and the suppression of radiocesium uptake (Matsuoka et al. 2019). Radiocesium transfer dynamics within the soil are influenced by the proportions of clay, sand, and organic matter in the soil (Hiraoka et al. 2015), so there is a need for longitudinal studies on the dynamics of radiocesium transfer to the lower layers.

11.4 Development of Techniques to Reduce Radiocesium Content in Trees and Orchards

11.4.1 Time-Course of Radiocesium Content in Mature Peaches Using Fruit Thinning

From the second half of 2011 until 2012, various measures were taken (including high-pressure washing of trees), but many issues could not be clarified until the trees had grown. In a study of cherries contaminated in the Chernobyl disaster, the radiocesium concentration in the fruit was found to have fallen to approximately one-third by the year after the disaster (Antonopoulos-Domis et al. 1996), and it is thought that there was probably a similar tendency after the FDNPP disaster. However, the actual concentrations in fruit were uncertain. The year after the disaster, the provisional regulation value for radiocesium concentration in food was changed from 500 to 100 Bq kg^{-1}, and in connection with that change, there was a demand for establishment of a more precise method for tracking the concentration in fruit. In this context, a study was performed that focused primarily on the changes in concentration in fruit from year to year during the growing season (Takata et al. 2014).

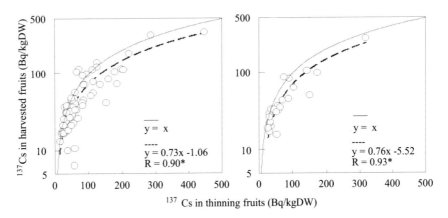

Fig. 11.3 Relationship between thinning fruit and harvested fruits in ^{137}Cs (from Takata 2016) left: each tree, $n = 70$, right: each orchard, $n = 24$). Dashed line means correlation line. Solid line means $Y = X$ line

The radiocesium concentration in fruit was generally found to reach a minimum level during the second growing season. The concentration in fruits during the growing season and at harvest was investigated and the variability in the measurement times and the feasibility of moving them forward were predicted (Fig. 11.3). This method was shown to be useful for the removal of contaminated trees. The peach fruit radiocesium concentrations were generally below the specified values in the fruit-producing region of Fukushima Prefecture after the FDNPP disaster, so this prediction system was not needed for peaches, whereas in persimmons (for making anpo-gaki), screening tests were performed to identify orchards where the harvested fruit had high radiocesium concentrations. This prediction method will be useful for radiocesium concentrations in fruit if any more nuclear disasters occur, either in Japan or overseas. In the case of deciduous fruit trees, the trees had no leaves and only the radiocesium fallout could be measured. In future, it will be necessary to elucidate transfer dynamics over a wide range of seasons.

11.4.2 Reducing Fruit Radiocesium Concentrations by Bark Washing

With fruit trees, bark peelability varies between species, with the bark being peeled readily from trees such as pear, grape, persimmon, and apple. By removing the rough bark, at least 80% of the radiocesium is removed (Sato et al. 2015). However, trees in which the seeds are hard stones, such as peaches and ume, have bark that is of low peelability. In this context, Abe et al. (2014) found that high-pressure washing of the bark of akatsuki peach trees, in July 2011, reduced the total radiation level by 55.9%. Washing was, however, not found to be effective for reducing radiocesium concentrations in fruit grown in 2011. This confirmed that most of the radiocesium was transferred internally within 60 days of contamination, suggesting that decontamination in July, approximately 4 months after the disaster, was already too late to restrict transfer of radiocesium into the tree.

In peaches and persimmons, washing the bark from winter 2011 to spring 2012 substantially reduced the radiocesium concentration of fruit harvested in 2012 in comparison with the fruit from unwashed trees (Abe et al. 2014; Sato 2014). This suggests that, in the case of uncontaminated persimmon trees, radiocesium transfer internally from the bark continues the year after contamination. After 3 years of persimmon bark formation, strip-shaped fissures form, and water droplets often collect in the scars left after the fissures, probably leading to the growth of moss and lichen on the bark (Abe et al. 2014). This could influence the radiocesium contamination.

11.4.3 Reducing Transfer to Fruit by Pruning and Cutting the Trunk

In the second year after radioactive contamination, radiocesium inside the fruit tree is the principal source of transfer to fruit (Antonopoulos-Domis et al. 1990), and pruning is considered to be effective for removing radiocesium that has been transferred internally into the tree. The effectiveness of pruning for removing radiocesium inside the tree is suggested to depend on not only the proportional distribution of radiocesium to the part that is removed, but also the amount of fruit collected (Hiraoka et al. 2015; Pröhl et al. 2003). However, in studies performed before the FDNPP disaster, the relationship between pruning and fruit radiocesium concentration had not been investigated.

In order to elucidate the relationship between pruning intensity and radiocesium concentration in chestnuts, Matsuoka et al. (2016) used 7-year-old trees and treatments of no-pruning, standard-pruning, and intense-pruning areas of the orchard in a March 2013 trial. In addition to measurement of the fruit radiocesium concentration over 2 years, in the second year, the trees were cut up, and the proportional radiocesium and radiopotassium distribution in each part was investigated. No significant differences in radiocesium concentration were found in the fruit that grew during the year of treatment, but the level of transfer of radiocesium to all fruit was substantially higher in the nonpruned area. The proportional distribution of radiopotassium to branches and roots that grew during the first to fourth years after the disaster was higher than that of radiocesium, suggesting that the radiocesium in the new growth had not come from the soil.

In March 2014, Kuwana et al. (2017), using persimmon trees of the hachiya variety (the fruit used to make anpo-gaki), included the following treatments: (1) no pruning; (2) standard pruning; (3) intense pruning; and (4) trunk-cutting, with the trunk cut at a height of 60–100 cm above the ground. The radiocesium concentrations were then measured in the fruit collected in each harvest season. The fruit radiocesium concentration decreased with time, except that in the no-pruning area in the fifth year, it was slightly higher than in the fourth year. No significant differences due to treatment mode were found in any of the study years, but the fruit radiocesium concentrations in the three pruned areas tended to show less variability than in the no-pruning areas.

In summary, although intense pruning is not clearly effective in reducing fruit radiocesium concentration, it is effective in restricting the variability of contamination strength in tree bodies and cultivation plots, and, in terms of ensuring yield, pruning operations are considered to be important for safe and stable fruit production.

11.4.4 Soil Radiocesium Elimination and Aerial Radiation Dose Reduction by Topsoil Stripping

Due to the deposition of radiocesium at high concentrations in the soil surface layer in orchards, the aerial radiation dose is higher in orchards than in surrounding areas, and there are concerns about workers suffering external exposure. In orchards, decontamination operations involving topsoil stripping have been performed, but many of the orchards in Fukushima Prefecture are on sloping land because it is a mountainous region, and there are limitations to the use of heavy machinery in such areas. Methods for topsoil stripping that do not involve use of machinery, using revegetation nets and other surface-covering materials, have been investigated (Sato et al. 2019). Approaches that make use of the ecological systems that exist in orchards are considered to be highly practical as orchard radiation decontamination methods.

11.5 Progress Toward Restarting Anpo-Gaki Shipment: Persimmons for Making Anpo-Gaki and Product Control

11.5.1 Restarting Shipments of Persimmons for Anpo-Gaki

In the northern part of Fukushima Prefecture, which is an important area for anpo-gaki production, immature fruit was sampled and tested from all persimmon cultivation plots after the FDNPP disaster. Screening of processed anpo-gaki for transport was performed before shipment. However, several years after the disaster, some products were confirmed to exceed the screening level, albeit only slightly (Table 11.1; Fukushima Prefecture Anpo-gaki Production Area Promotion Association 2018a), but Good Agricultural Practice (GAP) had been introduced in

Table 11.1 Results of radioactive substance inspection of Anpo-gaki (sulfur-smoked semi-dry persimmon)

Inspection year	Number of bags by concentration of Cs-137		
	<25 Bq kg^{-1}	25–50 Bq kg^{-1}	50 Bq kg^{-1}
2013	720,318 (90.80%)[a]	71,868 (9.06%)	1110 (0.14%)
2014	1,910,755 (94.54%)	107,097 (5.30%)	3148 (0.16%)
2015	3,470,991 (95.27%)	162,983 (4.47%)	9178 (0.25%)
2016	4,382,443 (98.29%)	72,917 (1.64%)	3208 (0.07%)
2017[b]	4,840,599 (99.96%)		1919 (0.04%)
2018[b]	5,292,949 (99.98%)		1144 (0.02%)

These data refers to the Fukushima prefecture anpo-gaki production area promotion association
[a]The numbers in parentheses are the ratio to the total number of bags inspected
[b]In 2017 and 2018, the number of bags is counted below 50 Bq kg^{-1}

Table 11.2 Cs-137 concentration of the young persimmon fruit [reorganized the data of Sekizawa et al. (2019)]

Survey site	Number of bags by concentration of Cs-137				
	<5 Bq kg^{-1}	5–10 Bq kg^{-1}	10–25 Bq kg^{-1}	25–50 Bq kg^{-1}	<50 Bq kg^{-1}
Orchard A	49 (67.1%)[a]	17 (23.2%)	7 (9.7%)	0 (0.0%)	0 (0.0%)
Orchard B	140 (84.8%)	16 (9.7%)	5 (3.6%)	2 (1.2%)	2 (1.2%)
Orchard C	2 (2.9%)	27 (39.1%)	34 (49.3%)	5 (7.2%)	1 (1.4%)
Orchard D	58 (71.6%)	20 (24.6%)	0 (0.0%)	0 (0.0%)	3 (3.7%)

[a]The numbers in parentheses are the ratio to the total trees

connection with processing (Fukushima Prefecture Anpo-gaki Production Area Promotion Association 2018b), so the adhesion potential to the external fruit during processing was reduced. It is highly likely, therefore, that the cause of the screening level being exceeded lay with the fruit used to make the anpo-gaki. The testing of immature fruit in the cultivation plots was on a sampling basis, so, if certain trees had higher contamination levels, the screening level could be exceeded in the fruit. Therefore, Sekizawa et al. (2019) investigated the intertree differences in fruit ^{137}Cs concentration in the same plots. In the anpo-gaki persimmons with excessive ^{137}Cs concentrations, the trees had high ^{137}Cs concentrations (Table 11.2).

11.5.2   Anpo-Gaki Product Control

To enable stable shipments of anpo-gaki, controls are required in both the production process for the fruit used, and in anpo-gaki processing for concentration variability in the fruit used and fluctuations in radiocesium concentration during processing. The present authors have verified the variability in radiocesium concentration in the fruit used, the proportional changes in radiocesium concentration during processing, and the risk of radiocesium contamination in commercial premises for anpo-gaki to be considered in risk management during anpo-gaki processing. The nondestructive radiation measurement devices used in full-quantity tests of anpo-gaki before shipment were strictly calibrated using model samples with known concentrations of certified reference materials in brown rice.

To clarify the variability in radiocesium concentration in persimmons used for anpo-gaki from the same tree, Sekizawa et al. (2019) performed a follow-up study of the differences in radiocesium concentration between persimmons collected from the same branches of the same trees, in northern Fukushima Prefecture. In 2013, up to 3.6-fold difference in concentration between persimmons from the same main branch was observed, and in 2014, when the radiocesium concentration had halved from the 2013 level, the concentration difference in the same main branch was up to approximately twofold. Both the concentration and concentration difference had decreased in the same main branch over time.

The commercial anpo-gaki processing facilities where persimmons were dried for an extended period with natural air flow were inspected, and radiocesium adhesion to the shelves and floor inside the facilities was confirmed. Although it was shown that the contamination status in the processing facilities had no effect on the anpo-gaki radiocesium concentration, the contamination risk is considered to be higher when persimmons are in direct contact with the floor and/or pipes, and decontamination must therefore be performed every year. For this reason, in 2014, preuse cleaning was included in the Fukushima Prefecture Anpo-gaki Production Area Promotion Association's "Manual of Good Agricultural Practice for Safe Anpo-gaki Production" (referred to below as the "GAP Manual").

Screening via nondestructive, full-quantity testing was introduced when anpo-gaki shipments restarted. In the first year after restarting anpo-gaki shipments, tests were performed on the basis of the product packaging as one box containing eight trays, each with six anpo-gaki. Therefore, for the testing device (separate from the calibration source), a reference model sample was prepared in accordance with the preparation method for the brown rice certified reference materials for radiocesium analysis (U-8 containers packed with 81 g of brown rice sample; Hachinohe et al. 2016). These samples were placed in an anpo-gaki tray, with the containers laid on their sides with three next to each other. The mass of each container was approximately twice that of the anpo-gaki; the brown rice samples had a ^{40}K content of 70–100 Bq kg^{-1}, which was approximately 50% that of the edible part of an anpo-gaki (174 Bq kg^{-1}). Using homogeneous brown rice samples with radiocesium concentrations of 0–140 Bq kg^{-1}, reference materials were prepared at five concentrations. These reference materials were introduced in the anpo-gaki full-quantity testing device and were used for verification of screening performance.

In summary, the radiocesium concentration in fruit used for anpo-gaki has a major effect on the concentration in the final product. Therefore, the risk of contamination in the commercial premises should be decreased by predicting the proportional increase in radiocesium concentration due to anpo-gaki processing. Radiocesium testing before and after anpo-gaki processing is effective for managing the risk in the processed product.

11.6 Societal Implementation of Study Results

The radiocesium decontamination measures for orchards are reflected in the funda-mental policies for decontamination of agricultural and forestry areas in Fukushima Prefecture and include washing trees, removing rough bark, and pruning. Data and statement methods supporting the effectiveness of such techniques were included in the "Policies for decontamination and technical measures relating to radiocesium in agricultural products" (Fukushima Prefectural Agriculture, Forestry and Fisheries Dept. 2012).

From December 2011 to March 2012, on the basis of the fundamental decon-tamination policies; in collaboration with prefectural, city, town, and village gov-ernments, and the Japan Agricultural Cooperatives, approximately 4300 ha of orchards of peaches, persimmons, apples, pears, and grapes included measures such as high-pressure washing of the trees (Fukushima Prefectural Agriculture, Forestry and Fisheries Dept. 2013). The outcome of these measures was that, since 2012, no restriction of shipment of the principal tree crops produced in Fukushima Prefecture, including peaches, apples, pears, and grapes, has been in effect.

In 2013, the third year of self-imposed control of anpo-gaki shipment, a model area for restarting processing was established, and a plan was established for shipment only of products confirmed to be safe using a nondestructive testing device. To establish this model area, the radiocesium concentration of the persim-mons used for anpo-gaki had to meet specified values. Before using the nondestruc-tive testing device, radiocesium reference samples were used in performance tests on the device for calibration.

The results of studies on the prevention of secondary contamination during cultivation and processing are reflected in the GAP Manual (Fukushima Prefecture Anpo-gaki Production Area Promotion Association 2018b), and all producers in the anpo-gaki production process comply with this. Cleaning of commercial premises is performed as a cooperative operation, maintaining a consistent operational level, on the basis of the GAP Manual.

It is expected that replacement of trees that still have fruit radiocesium concen-trations exceeding the specified value will be performed after discussing and decid-ing upon appropriate compensation.

All data obtained in an agriculture, forestry, and fisheries survey, which covers fruit and is performed as emergency monitoring of environmental radiation, is made available on the homepage of the Fukushima Association for Securing Safety of Agricultural Products. In addition, all test data for anpo-gaki obtained using the nondestructive testing device are made available on the homepage of the Fukushima Division of the National Federation of Agricultural Cooperative Associations. In parallel with efforts to ensure safety at production sites, information disclosure and the expansion of public relations activities has ensured that there has been a recovery in demand. The value of prefectural fruit shipment, which in 2011 had fallen to ¥19.7 billion (63% of that in 2010), had recovered to ¥27.1 billion by 2016 (Ministry of Agriculture, Forestry and Fisheries (Japan), 2010, 2011, 2016). In the case of anpo-gaki, the shipments increased yearly since processing was restarted, so that in 2017, shipments were approximately 76% of the 2010 level.

The trade in fruit trees in Fukushima Prefecture is an important industry, and can be reflected in the status of peach trees. Figure 11.4a shows the mean price at the central wholesale market for peaches produced in Fukushima Prefecture. The mean price in the prefecture was ¥438 in 2010, before the disaster, but it decreased to ¥222 in the year of the disaster. Subsequently, as recovery continued, it reached ¥429 in 2015, which was equal to the 4-year predisaster mean. In 2018 and 2019, the mean sale price was approximately ¥500, which was even higher than before the disaster. However, this interpretation may be subjective, and in order to explain this, Fig. Y has been prepared as a follow-up of the national mean sale price shown in Fig. 11.4b. When the mean sale prices of peaches in Fukushima Prefecture and the whole of Japan are compared, the FDNPP disaster caused sales and economic problems. The price of peaches in Fukushima Prefecture before the disaster was ¥20–¥50 below the national mean. This price differential was linked to the production volume and marketing routes in Fukushima Prefecture, although no major fluctuation in that respect is shown here. In contrast to prices only within Fukushima Prefecture, in 2018 and 2019, the mean price of peaches produced in Fukushima Prefecture was nearly ¥100 less than the national mean. Over the past 10 years, Japanese peach production has been supported by the national policy of establishing high prices and high demand, especially in relation to overseas exports, and production can be seen to have increased markedly as a result. In the context of this expansion of the industry itself, production in Fukushima Prefecture has been unable to ride the tide of high-price sales, such that the difference from the national mean has widened as the latter has increased. The period from 2010 to 2020 was one of marked growth in Japanese peach production, but in Fukushima Prefecture, it was a period of recovery from the earthquake-related disaster, and the opportunity to achieve high-price sales was thus missed. This "lost decade" due to the FDNPP disaster means that the earthquake had a major negative impact on fruit production in Fukushima Prefecture.

11.7   Conclusions and Future Issues

The radiocesium concentration in fruit in the second year after the disaster was found to be approximately one-third of that during the first year, and by the third year, it had again fallen to a third; this rate of decrease is greater than that due to the half-life predicted by physics (Kusaba et al. 2015a, b; Renaud and Gonze 2014; Takata 2016, 2019; Yuda et al. 2014). These findings were similar to those reported from several studies on fruit trees after the Chernobyl disaster (Antonopoulos-Domis et al. 1990, 1996; Madoz-Escande et al. 2002; Mück 1997; Pröhl et al. 2006). However, there are interspecies differences in the pattern of radiocesium concentration decrease in fruit. In yuzu (a small, sour citrus fruit; *Citrus junos*) and persimmons, certain trees have especially high concentrations in the same orchards, but the cause of this has not

a

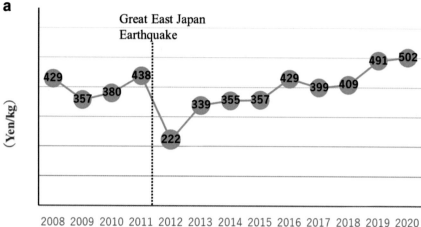

b

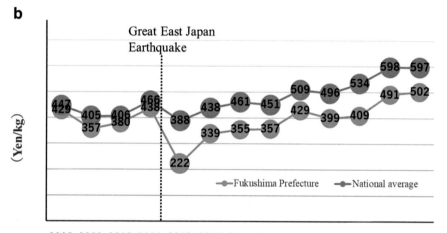

Fig. 11.4 (**a**) Changes in the average unit price of peaches produced in Fukushima Prefecture at the Tokyo Central Wholesale Market (From Ministry of Agriculture, Forestry and Fisheries (Japan) (2021)). (**b**) Changes in the average unit price of peaches produced in Fukushima Prefecture and national average at the Tokyo Central Wholesale Market (From Ministry of Agriculture, Forestry and Fisheries (Japan) (2021))

been sufficiently determined (Sekizawa et al. 2019). With respect to long-term changes in radiocesium concentration in fruit trees, there have been few follow-up studies at the same sites (Tagami 2016), and it is uncertain what changes will occur in future. In this respect, it is important not only to perform follow-up studies on concentrations in fruit, but also to ascertain the level of radiocesium, in terms of both concentration and absolute quantity, in the fruit trees from which fruit is taken. In previous studies on fruit trees, it has been shown that radiocesium in the soil makes almost no contribution to radiocesium in the fruit, and the following explanations for this have been put forward: (1) in Japan, even in annual crops, the coefficient of transfer to the fruit is low; (2) uptake of radiocesium deposited directly on fruit trees into the tree via the bark is not complete, and radiocesium in the tree makes a major contribution to the radiocesium concentration in the fruit (Takata et al. 2012c, 2013a); (3) radiocesium in the tree still remains there the following year, so that is the origin of radiocesium in the fruit (Takata et al. 2013a, b); and (4) radiocesium is currently distributed heterogeneously in the upper layers of the soil only; most fruit trees grown in Fukushima Prefecture (excluding blueberries) do not form their main root zones in the top 5 cm of soil, which would be responsible for the uptake of radiocesium (Kusaba et al. 2016; Sato et al. 2019; Takata et al. 2013b). Suggestion (4) is linked to the fact that fertilization with potassium is usually ineffective. It is therefore necessary to perform a follow-up study on the trees that produce radiocesium in the fruit. It is important to note that the accumulation of high levels of radioactive cesium in the upper soil layers of orchards is not only related to the safety of the agricultural products produced, but also to the safety of the growers' operations. In relation to yuzu trees growing in the lower parts of the slopes, it is suspected that radiocesium was absorbed via the roots due to flooding caused by heavy rain (Sato et al. 2019). It is necessary to perform long-term monitoring of the potential for radiocesium uptake via the roots. Little progress has been made with measures to decontaminate orchards by topsoil-stripping, and there are numerous orchards where the established guidelines (Ministry of the Environment (Japan) 2011) have still not been met (Sato et al. 2019). There is therefore also a need for long-term, continuous measurement of the aerial radiation dose rate.

In Fukushima Prefecture, 8 years after the disaster, some fruit-growing regions are specified as evacuated areas, and measures to restart commercial agriculture in these areas have yet to be established. In areas where almost no radioactive nuclei have been detected, problems in terms of consumption behavior remain. With no direct relationships to radioactive nuclei, this has been linked to the phenomenon of "misinformation." A major factor involved in this phenomenon is considered to be the delays in updating information for consumers. As a result, even when accurate data is obtained, neither the environment of the sender nor the receiver of that information is satisfactory. In the Tokyo region, which is a major consumption area, opportunities to transmit information about fruit radiocesium concentrations are much fewer than in Fukushima Prefecture. The current situation is disheartening, as straightforward, positive information about areas where cultivation is currently possible and acceptable, which is gradually being accumulated and collated, does not catch people's attention as much as sensational information about specified safety levels having been exceeded. In addition, in terms of the positioning of fruit in the Japanese market, one difference from other products is the strong perception of fruit being sold as gifts. There are differences between consumption behavior relating to items for gifts and for use in one's own home. For example, once an alternative avenue for gifts has been identified, it is difficult for the original production site to win those consumers back. Long-term monitoring is as important as the safety survey soon after the disaster. Scientific issues pertaining to this problem must be investigated continuously in 10-year intervals, and the aspect of this being a *scientific* study must be emphasized. The study must also be accompanied by firm and clear statements that do not misconstrue concealed safety concerns.

The fruit tree industry does not depend solely on production. Making progress with a wide range of research, from studies performed close to the production site with the aim of reducing radiocesium levels to analysis of the consumption psychology of the actual purchasers, should provide a boost that will lead to sales price recovery.

References

Abe K, Sato M, Takita K, Yuda M, Ajito Y, Ono T, Kikunaga H, Otsuki T, Muramatsu Y (2014) Effect of aging decrement and decontamination trial to radiocesium concentration in fruits and leaves of the Japanese persimmon contaminated at the dormant periods. Bulletin of the Fukushima Agricultural Technology Centre. Radioactive materials. Special issue: 74–77

Antonopoulos-Domis M, Clouvas A, Gagianas A (1990) Compartment model for long-term contamination prediction in deciduous fruit trees after a nuclear accident. Health Phys 58: 737–741

Antonopoulos-Domis M, Clouvas A, Gagianas A (1996) Long term radiocesium contamination of fruit trees following the Chernobyl accident. Health Phys 71:910–914

Aoki D, Asai R, Tomioka R, Matsushita Y, Asakura H, Tabuchi M, Fukushima K (2017) Translocation of ^{133}Cs administered to *Cryptomeria japonica* wood. Sci Total Environ 584: 88–95

Carini F, Lombi E (1997) Foliar and soil uptake of ^{134}Cs and ^{85}Sr by grape vines. Sci Total Environ 207:157–164

Fromm J (2010) Wood formation of trees in relation to potassium and calcium nutrition. Tree Phys 30:1140–1147

Fukushima Prefectural Agriculture, Forestry and Fisheries Dept. (2012) Guidelines for decontamination and technical measures related to measures against radioactive cesium in agricultural products. http://www.pref.fukushima.lg.jp/uploaded/attachment/61508.pdf

Fukushima Prefectural Agriculture, Forestry and Fisheries Dept. (2013) Record of the Great East Japan Earthquake in the field of agriculture, forestry and fisheries (1st edition), pp 104–105

Fukushima Prefecture Anpo-gaki Production Area Promotion Association (2018a) Radioactive substance inspection information of 2017 Anpo persimmon. http://www.fs.zennoh.or.jp/product/vegi/furuit/winter/pdf/kensa20180404.pdf

Fukushima Prefecture Anpo-gaki Production Area Promotion Association (2018b) Agricultural production process management (GAP) practice manual for safe Anpo-gaki production (2018 revised edition), pp 1–8

Hachinohe M, Todoroki S, Unno Y, Miura T, Yunoki A, Hamamatsu S (2016) Preparation of certified reference materials for radioactivity measurement using brown rice. Radioisotopes 65:169–180

Hamada N, Ogino H, Fujimichi Y (2012) Safety regulations of food and water implemented in the first year following the Fukushima nuclear accident. J Radiat Res 53:641–671

Hiraoka K, Matsuoka K, Kusaba S (2015) Differences in the distribution of radiocesium in deciduous and evergreen fruit trees: a case study. Soil Sci Plant Nutr 61:230–234

IAEA (2003) Modeling the transfer of radionuclides to fruit. Report of the fruits working group of BIOMASS Theme3. IAEA, Vienna, p 151

IAEA (2010) Handbook of parameter values for the prediction of radionuclide transfer in terrestrial and freshwater environments technical reports series, No. 472. IAEA, Vienna

Islam MA, Begum S (2012) Histochemical and anatomical studies of phloem and xylem cells of jackfruit (*Artocarpus heterophyllus* lam.) tree. Int J Nat Sci 2:1–7

Iwabuchi K (2014) Inhibition of radiocesium absorption in a blueberry. Bulletin of the Fukushima Agricultural Technology Centre. Radioactive materials. Special issue: 82–85

Komatsu T (2014a) Realities and support policies regarding countermeasures for radioactive materials in nuclear disaster areas. J.J. Farm Manage 32:25–35

Komatsu T (2014b) Actual condition of sale and the trend of direct selling by fruit farm management after a nuclear hazard. J.J. Farm Manage 32:47–52

Kusaba S, Matsuoka K, Abe K, Ajito H, Abe M, Kihou N, Hiraoka K (2015a) Changes in radiocesium concentration in a blueberry (*Vaccinium virgatum* Aiton) orchard resulting from radioactive fallout soil. Sci Plant Nutr 61:169–173

Kusaba S, Matsuoka K, Saito T, Kihou N, Hiraoka K (2015b) Changes in radiocesium concentration in a Japanese chestnut (*Castanea crenata* Sieold & Zucc.) orchard following radioactive fallout. Soil Sci Plant Nutr 61:165–168

Kusaba S, Matsuoka K, Abe K, Ajito H, Abe M, Sakuma N, Saito Y, Shimura H, Kihou N, Hiraoka K (2016) Effect of soil surface management on radiocesium concentrations in apple orchard and fruit. Hort J 85:30–36

Kuwana A, Adachi Y, Mastumoto N, Satio M (2017) Effect of pruning and planting on the temporal change of radiocesium activity concentration in Japanese persimmon 'Hachiya'. Hort Res 16(2): 167

Madoz-Escande C, Colle C, Adam C (2002) Evolution of cesium and strontium contamination deposited on vines. Radioprotection 37(C1):515–520

Mahara Y, Ohta T, Ogawa H, Kumata A (2017) Atmospheric direct uptake and long-term fate of radiocesium in trees after the Fukushima nuclear accident. Sci Rep 4:7121. https://doi.org/10.1038/srep07121

Matsuoka K, Kusaba S, Nishio S, Hiraoka K (2016) Effects of winter pruning on the concentration and amount of radiocaesium in Japanese chestnut trees after radionuclide deposition. Radioisotopes 65:367–376

Matsuoka K, Moritsuka N, Kusaba S, Hiraoka K (2019) Concentrations of natural stable Cs in organs of blueberry bushes grown in three types of soils treated with acidification or fertilization. Hort J 88:31–40

Ministry of Agriculture, Forestry and Fisheries (Japan) (2012) Research books of dry-persimmon production. http://www.maff.go.jp/j/tokei/kouhyou/tokusan_kazyu/index.html

Ministry of Agriculture, Forestry and Fisheries (Japan) (2018) Statistical yearbook of ministry of agriculture, forestry and fisheries, 2010, 2011, 2016

Ministry of Agriculture, Forestry and Fisheries (Japan) (2021) Survey on the actual distribution of agricultural products produced in Fukushima Prefecture in the first year of the Ordinance

Ministry of the Environment (Japan) (2011) Guidelines for measures such as decontamination. https://www.jcmanet.or.jp/daishinsai/2011/kankyo_guideline/02.pdf

Mio S, Matsumoto S (1979) Morphological observation on intercellular spaces in wood ray. Bull Kyushu Univ 51:1–12

Mück K (1997) Long-term effective decrease of cesium concentration in foodstuffs after nuclear fallout. Health Phys 72:659–673

Pfautsch S, Renard J, Tjeolker MG, Salih A (2015) Phloem as capacitor: radial transfer of water into xylem of tree stems occurs via symplastic transport in ray parenchyma. Plant Phys 167:963–971

Pröhl G, Fiedler I, Koch-Steindl H, Leser C, Treutter D (2003) Interzeption und Translokation von Radiocäsium bei Obst und Beeren. Bonn, Bundesministerium für Umwelt, Naturschutz und Reaktorsicherheit

Pröhl G, Ehlken S, Fiedler I, Kirchner G, Klemt E, Zibold G (2006) Ecological half-lives of ^{90}Sr and ^{137}Cs in terrestrial and aquatic ecosystems. J Environ Radioact 91:41–72

Renaud P, Gonze MA (2014) Lessons from the Fukushima-and Chernobyl accidents concerning the ^{137}Cs contamination of orchard fresh fruits. Radioprotection 49:169–175

Sato M, Abe K, Takata D, Ono T, Takase T, Kawatsu K, Tanoi K (2014) Investigation of the radiocesium migration pathway into the deciduous fruit tree contaminated at the dormant period. Bulletin of the Fukushima Agricultural Technology Centre. Radioactive materials. Special issue 70–73

Sato M, Takata D, Tanoi K, Ohtsuki T, Muramatsu Y (2015) Radiocaesium transfer into the fruit of deciduous fruit trees contaminated during dormancy. Soil Sci Plant Nutr 61:156–164

Sato M, Akai H, Saito Y, Takase T, Kikunaga H, Sekiya N, Ohtsuki T, Yamaguchi K (2019) Use of different surface covering materials to enhance removal of radiocaesium in plants and upper soil from orchards in Fukushima prefecture. J Environ Radioact 196:204–211

Sekizawa H, Horie Y, Kuwana A, Hachinohe M, Hamamatsu S (2019) Differences in radioactive caesium-137 concentrations in young persimmon fruit from trees in the same orchard. Radio-isotopes 68:345–354

Suzuki Y, Sato M, Yanai K (2018) Efforts to restore the environment in Fukushima (9th) Countermeasures for radioactive substances in agricultural products. Atomic Energy Soc Jpn (AESJ) 60:41–45. (In Japanese)

Tagami K (2016) Effective half-lives of radiocesium in terrestrial plants observed after nuclear power plant accidents. In: Gupta DK, Walther C (eds) Impact of cesium on plants and the environment. Springer International Publishing, Cham, pp 125–138

Takata D (2013) Distribution of radiocaesium from the radioactive fallout in fruit trees. In: Nakanishi TM, Tanoi K (eds) Agricultural implications of the Fukushima nuclear accident. Springer, Tokyo, pp 143–162

Takata D (2016) Translocation of radiocesium in fruit trees. In: Nakanishi TM, Tanoi K (eds) Agricultural implications of the Fukushima nuclear accident: the first 3 years. Springer Japan, Tokyo, pp 119–143

Takata D (2019) The transition of radiocesium in peach trees after the Fukushima nuclear accident. In: Nakanishi TM, O'Brien M, Tanoi K (eds) Agricultural implications of the Fukushima nuclear accident (III): after 7 years. Springer Japan, Tokyo, pp 85–105

Takata D, Fukuda F, Kubota N (2008) Effects of different cultural practice on the occurrence of reddish-pulp fruit and fruit development in 'Beni Shimizu' peach. Hort Res 7:367–373

Takata D, Yasunaga E, Tanoi K, Nakanishi T, Sasaki H, Oshita S (2012a) Radioactivity distribution of the fruit trees ascribable to radioactive fall out (II): transfer of radiocaesium from soil in 2011 when Fukushima Daiichi nuclear power plant accident happened. Radioisotopes 61:517–521

Takata D, Yasunaga E, Tanoi K, Kobayashi N, Nakanishi T, Sasaki H, Oshita S (2012b) Radio-activity distribution of the fruit trees ascribable to radioactive fall out (III): a study on peach and grape cultivated in South Fukushima. Radioisotopes 61:601–606

Takata D, Yasunaga E, Tanoi K, Nakanishi T, Sasaki H, Oshita S (2012c) Radioactivity distribution of the fruit trees ascribable to radioactive fall out: a study on stone fruits cultivated in low level radioactivity region. Radioisotopes 61:321–326

Takata D, Yasunaga E, Tanoi K, Nakanishi T, Sasaki H, Oshita S (2012d) Radioactivity distribution of the fruit trees ascribable to radioactive fall out (IV): cesium content and its distribution in peach trees. Radioisotopes 61:607–612

Takata D, Sato M, Abe K, Yasunaga E, Tanoi K (2013a) Radioactivity distribution of the fruit trees ascribable to radioactive fall out (V): shifts of caesium-137 from scion to other organs in 'Kyoho' grapes. Radioisotopes 61:455–459

Takata D, Sato M, Abe K, Yasunaga E, Tanoi K (2013b) Radioactivity distribution of the fruit trees ascribable to radioactive fall out (VI): effect of heterogeneity of caesium-137 concentration in soil on transferability to grapes and fig trees. Radioisotopes 62:533–538

Takata D, Sato M, Abe K, Tanoi K, Kobayashi N, Yasunaga E (2014) Shift of radiocaesium derived from Fukushima Daiichi nuclear power plant accident in the following year in peach trees. In: 29th International horticultural congress, impact of Asia-Pacific horticulture, vol 117, p 215

Van Bell AJE (1990) Xylem–phloem exchange via the rays: the undervalued route of transport. J Exp Bot 41:631–644

Yuda M, Sato M, Abe K, Nukada M, Saito Y, Yamaguchi N, Takita K, Ajito Y, Ono T, Muramatsu Y, Ho K, Otsuki T (2014) Temporal changes of the radiocesium concentrations in different parts of deciduous fruit trees and effect of decontamination trial to bark with high pressure washers. Bulletin of the Fukushima Agricultural Technology Centre. Radioactive materials. Special issue 78–81

Chapter 12
Overview of Radiocesium Dynamics in Forests: First Decade and Future Perspectives

Shoji Hashimoto

12.1 Introduction

Fukushima Prefecture and the surrounding areas are among the most forested areas in Japan (Fig. 12.1); ~70% of the contaminated areas are covered with forests (Hashimoto et al. 2012). Accordingly, although the amount of total deposition on the land is still in debate, ~70% of it was deposited onto forest ecosystems (Kato et al. 2019a). Therefore, the forest is one of the key ecosystems in the radioactive contamination caused by the Fukushima Daiichi Nuclear Power Plant accident.

In Fukushima, there are 0.97 million ha of forests. The forests in Fukushima and adjacent regions are rich in tree species, like the forests in other areas of Japan. The most dominant species are the Japanese cedar (*Cryptomeria japonica*), the Japanese cypress (*Chamaecyparis obtusa*), the red pine (*Pinus densiflora*), and the Konara oak (*Quercus serrata*). These trees in artificial forests are used for building, paper, and mushroom cultivation. Furthermore, in forests, local inhabitants collect wild edible mushrooms and vegetables, and hunt wild game for meat (e.g., sika deer, Asian black bear, and wild boar). Therefore, the radioactive contamination by the Fukushima Daiichi Nuclear Power Plant accident seriously affected the lives of the local inhabitants.

Therefore, the radiocesium dynamics in the forest are of great concern to people and the authorities. The studies conducted after the Chernobyl (Chornobyl) Nuclear Power Plant accident have reported that the radiocesium deposited onto forests migrates within forest ecosystems (IAEA 2006; Shaw 2007). The radiocesium is

S. Hashimoto (✉)
Laboratory of Silviculture, Graduate School of Agricultural and Life Sciences, The University of Tokyo, Tokyo, Japan

Department of Forest Soils/Center for Biodiversity and Climate Change, Forestry and Forest Products Research Institute (FFPRI), Tsukuba, Ibaraki, Japan
e-mail: a-shoji.hashimoto@g.ecc.u-tokyo.ac.jp; shojih@ffpri.affrc.go.jp

© The Author(s) 2023
T. M. Nakanishi, K. Tanoi (eds.), *Agricultural Implications of Fukushima Nuclear Accident (IV)*, https://doi.org/10.1007/978-981-19-9361-9_12

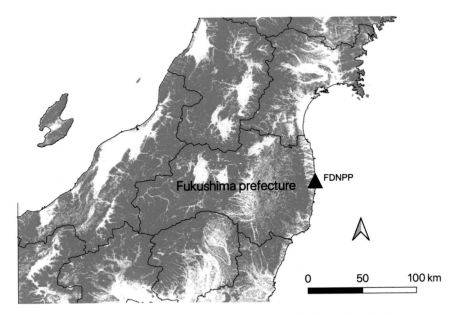

Fig. 12.1 Spatial distribution of forests in Fukushima area (Ministry of Land, Infrastructure, Transport and Tourism 2020). Black triangle indicates the location of the Fukushima Daiichi Nuclear Power Plant

first trapped by the tree canopy and migrates to the forest floor afterward. The mineral soil under the organic layer of the soil surface becomes the largest reservoir of radiocesium eventually. With time, spanning from days to years, migrations occur and are driven by various physical and biological processes in forest ecosystems. However, the overall picture described above is not necessarily inferred from the abundance of observation data (particularly limited data in the first few years).

In Fukushima, immediately after the fallout, many researchers and governmental authorities collected huge amounts of data that described the spatiotemporal radiocesium dynamics. This chapter describes the overall radiocesium dynamics in forest ecosystems that were revealed by various research studies conducted in the first decade after the accident. Also, it discusses the future status of forest radiocesium with predictions from modeling studies.

12.2 Decrease in Radiation Dose in Forests

The important characteristic of radioactive contamination is its radioactive decay. Figure 12.2 shows the temporal changes in the air dose rates measured in forests in Fukushima Prefecture (Fukushima Prefecture Forestation Division 2021). The long-term monitoring clearly illustrates the decreasing trend of the radiation dose in forests, which is consistent with the theoretical changes caused by the physical

Fig. □12.2 Temporal changes in air dose rate observed at 362 forest sites in Fukushima Prefecture (solid circle) (Fukushima Prefecture Forestation Division 2021). The line shows the theoretical curve based on the decay of ^{134}Cs and ^{137}Cs

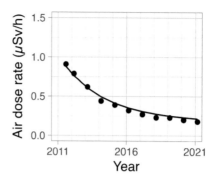

decay of ^{134}Cs and ^{137}Cs. As in other ecosystems, the decrement in the air dose rate is more rapid in the first several years due to the decay of ^{134}Cs. In the later phase of the first decade after the accident, the decrement slows down, because most of the remaining radiocesium is ^{137}Cs, which has a half-life of 30 years. The decrease in the air dose rate may occur faster than predicted based on the theoretical curve due to the further migration of radiocesium in the deeper layers of mineral soils. However, the decrease would basically follow the theoretical curve, because the migration of radiocesium within the soil is very slow, and most radiocesium will stay in the shallowest layer of forest soils. Besides, the forest retains most radiocesium without substantial outflow via stream flows. These characteristics of temporal changes in the air dose rate in forests are the fundamental fact for local people and authorities that take safe and effective countermeasures against contaminated forests.

12.3 Intensively Investigated Parts of Forests

The obtained forest radioactivity data have been published in scientific journals, government reports, and the Web. Published data on activity concentration and inventory of trees, soils, and mushrooms observed in forests were comprehensively compiled (Hashimoto et al. 2020b). The compilation showed that trees were investigated more intensively, but soils and mushrooms were also moderately investigated (Table 12.1). The inventory data for the soil were actively obtained, whereas those for the tree part were less investigated.

As expected, the compilation found much data for the Japanese cedar, because it is the most widely distributed tree species in managed forests (Table 12.2). However, the Konara oak, an important tree species for mushroom cultivation in Japan, was less studied.

Concerning the parts of the Japanese cedar, the leaf (needle) was mostly investigated intensively (Table 12.2). However, the woods (bulk wood, heartwood, and sapwood) were well studied, probably because wood is commercially the most important forest product. Interestingly, the pollen of the Japanese cedar was also

Table 12.1 Data count for trees, soils, and mushrooms (activity concentration and inventory) (Hashimoto et al. 2020b)

Sampling year	Activity concentration			Inventory	
	Tree	Soil	Mushroom	Tree	Soil
2011	578	1120	151	80	1001
2012	2070	851	333	76	527
2013	1148	417	508	120	381
2014	2172	622	592	84	616
2015	1376	591	632	99	569
2016	1071	420	566	4	369
2017	178	84	407	8	58
Total	8593	4105	3189	471	3521

Table 12.2 Counts for activity concentration data for trees by species and parts (Hashimoto et al. 2020b)

Part	Species					
	Cedar (*Cryptomeria japonica*)	Cypress (*Chamaecyparis obtusa*)	Pine (*Pinus densiflora*)	Oak (mainly *Quercus serrata*)	Other	Total
Leaf	1081	194	240	118	350	1983
Branch	251	109	149	172	105	786
Bark	360	117	185	151	31	844
Inner bark	82	36	22	28	20	188
Outer bark	416	120	252	38	21	847
Wood	267	85	126	45	5	528
Heartwood	609	182	328	193	34	1346
Sapwood	621	190	344	215	48	1418
Pollen	115	0	0	0	0	115
Other	290	2	3	216	27	538
Total	4092	1035	1649	1176	641	8593

well investigated, because it is the main cause of the hay fever in Japan, and the potential pollen contamination could be of great concern to the public.

12.4 Radiocesium Dynamics in Forests

12.4.1 Trees to Soils

Most of the radiocesium deposited onto forests was first trapped by the tree canopy (Kato et al. 2012, 2019a; Gonze et al. 2021). The reported interception rates vary between studies due to the different observation periods, methods, and forest types (Hashimoto et al. 2020c). Gonze et al. compiled the early data and back-calculated the interception rate for coniferous trees during deposition; the estimated value was

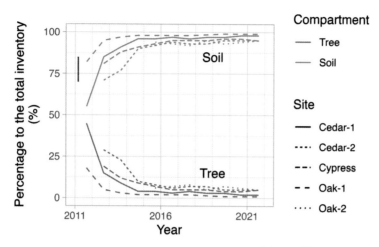

Fig. 12.3 Changes in percentages of radiocesium inventory (^{134}Cs + ^{137}Cs) in trees and soil (organic layer + mineral soil) to total inventory in the forest observed at long-term monitoring sites operated by FFPRI (Forestry Agency 2022). The red vertical line in 2011 is the tree interception ratio in March 2011 estimated by Gonze et al. (2021)

0.7–0.85 in March 2011 (Gonze et al. 2021). The interception ratio for the deciduous trees, which had no leaves in March, was lower than that for coniferous trees.

As reported in many studies, the radiocesium then rapidly moved from the tree canopy to the forest floor in time scales of days to years through litterfall and rainfall (Itoh et al. 2015; Kato et al. 2019b). Figure 12.3 shows the percentage of the radiocesium inventory in the tree canopy and soil (soil surface organic layer + mineral soil layer) relative to the total inventory observed in the forests in Fukushima Prefecture (Forestry Agency 2022). Note that the sampling was conducted during the summer yearly, and that the first data were taken 5 months after the fallout in 2011. Although the tree inventory ratio varies among sites and species, the tree inventory decreases with time, and accordingly, the soil inventory increases with time. Within a few years, the soil surface organic layer (A_0 layer) and mineral soil retain more than 90% of the total inventory in the forests. The activity concentration of ^{137}Cs in needles and branches exponentially decreased with time from 2011 to 2015, with effective ecological half-lives of 0.45–1.55 and 0.83–1.69 years for needles and branches, respectively (Imamura et al. 2017).

12.4.2 Soil

As shown above, the radiocesium trapped by the tree biomass moved to the forest floor (Fig. 12.3). However, most of the radiocesium deposited onto the forests is now in mineral soils, not in the soil surface organic layer (Fig. 12.4). The soil surface organic layer retains almost the same amount of inventory as in the mineral soil layer

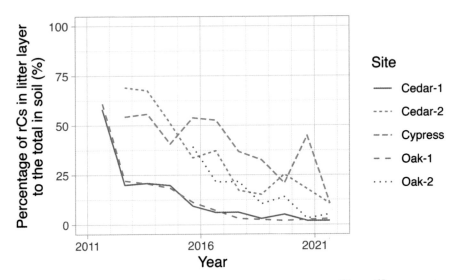

Fig. 12.4 Time dependency of percentage of radiocesium inventory (rCs: ^{134}Cs + ^{137}Cs) in the soil surface organic layer (litter layer, A_0) to total inventory in soil (soil surface organic layer + mineral soil layer) observed in long-term monitoring sites operated by the FFPRI (Forestry Agency 2022)

in the early stage, but the percentage continuously decreases over time (Fig. 12.4). Imamura et al. compiled the radiocesium distributions in the soil surface organic and mineral soil layers observed between 2011 and 2017, and found that the radiocesium fraction in the soil surface organic layer drastically decreased in the first 6 years from ~60% in 2011 to ~10% in 2017 (Imamura et al. 2017). Probably because the soil surface organic layer is thinner in Japan than in the forests in European countries affected by the Chernobyl accident, radiocesium did not remain long in the surface organic layer in Japan.

In the mineral soil layer, most radiocesium materials remain in the shallow layer (Fig. 12.5). Figure 12.5a shows the temporal changes in the activity concentration in each soil layer in a forest in Kawauchi village, Fukushima (Forestry Agency 2022). First, the activity concentration at depths of 0–5 cm increases from 2011 to 2012. Afterward, the highest concentration between layers is maintained. The vertical distribution of the radiocesium inventory observed in 2011 and 2015 is shown in Fig. 12.5b (Imamura et al. 2017). The radiocesium inventory in the surface layer has remained the largest since 2011.

12.4.3 Wood

The activity concentrations of wood, which is the most important part of a tree in terms of commercial use, differ among tree species and forests. Figure 12.6 shows the temporal trends in the activity concentrations of the Japanese cedar, Japanese cypress, red pine, and Konara oak observed in forests in Fukushima. These trends

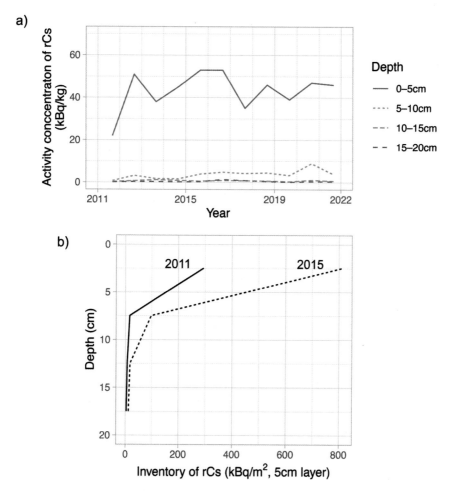

Fig. 12.5 Vertical distributions of radiocesium (rCs: [134]Cs + [137]Cs) in Japanese cedar forest in Kawauchi village (long-term monitoring sites of FFPRI). (**a**) Temporal changes in activity concentrations (Forestry Agency 2022), and (**b**) vertical distributions of radiocesium inventory observed in 2011 and 2015 (Imamura et al. 2017)

differ even for the same tree species. However, the Konara oak shows increasing trends, whereas the cedar shows stable or small increasing and decreasing trends. The normalized activity concentration, which is almost comparable to T_{ag}, mainly ranges from 10^{-4} to 10^{-3} (m²/kg, dry weight).

The T_{ag} values for the wood observed in Fukushima were collated and compared with those in Chernobyl studies (Hashimoto et al. 2020c). The range and geometric mean of the T_{ag} values for the conifer (pine excluded), pine, and oak are similar to those observed after the Chernobyl accident. Although the record number is small, those for the pine are probably lower than those reported for the Chernobyl accident, including the data measured for hydromorphic soils (Hashimoto et al. 2020c). The

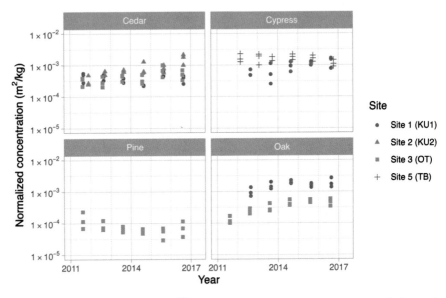

Fig. 12.6 Activity concentration of ^{137}Cs for Japanese cedar, Japanese cypress, red pine, and Konara oak normalized with total inventory of ^{137}Cs in 2012. [The data was originally from Ohashi et al. (2017), and were normalized by the author]

similarity indicates that the overall magnitude of transfer to the wood was similar and comparable, although the number of samples and different species and timing since the fallout must be considered. The comparison suggests the mutual global applicability of radioecological parameters and the need for further collation of parameters. However, Fig. 12.6 shows that the transfer to wood and its time dependence vary among species and sites, thereby emphasizing careful monitoring of trees and soils at affected areas where accidents occurred is always essential. The cause of the variability remains unclear; thus, future studies should be conducted.

12.4.4 Wild Food in Forests

The radiocesium deposited onto forests was also transferred to wild edible foods, such as wild mushrooms, wild edible plants, and game animals. The consumption of these contaminated foods in forests is an even more sensitive issue in affected areas. After the accident, the government and scientists have monitored the radioactivity levels in wild foods in the forests. The data revealed huge variations of radiocesium transfer among species (Tagami et al. 2016; Komatsu et al. 2019, 2021; Hashimoto et al. 2020c). For example, Komatsu analyzed open government monitoring data for wild mushrooms and found that the normalized concentration ranged from 1.1×10^{-4} to 2.3×10^{-2} (m^2/kg, fresh weight) and that the mycorrhizal species tend to have higher radioactive concentrations than saprophytic species (Komatsu et al.

Fig. 12.7 Violin plot of normalized concentration (*NC*: almost comparable to T_{ag}) for mushroom species according to ecological types. Concentration is normalized with total deposition (Komatsu et al. 2019)

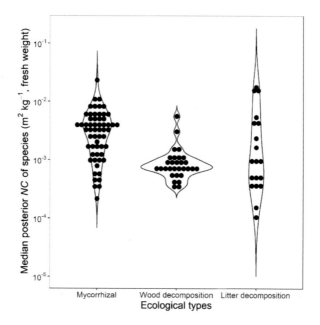

2019) (Fig. 12.7). It was also confirmed that wild plants generally had lower normalized concentrations than mushrooms, although some species of wild plants showed much higher transfer (e.g., koshiabura, *Eleutherococcus* (*Chengiopanax*) *sciadophylloides*) (Komatsu et al. 2019, 2021). A similar analysis was also conducted using the monitoring data for game meat (Tagami et al. 2016). The analysis of the data collected between 2011 and 2015 revealed that the geometric mean values of the T_{ag} values of ^{137}Cs in 2015 for the Asian black bear, wild boar, sika deer, and copper pheasant were of similar values, ranging from 1.9×10^{-3} to 5.1×10^{-3} (m^2/kg, fresh weight), whereas those for the green pheasant and wild duck were about one order of magnitude lower than these species, ranging from 1.0×10^{-4} to 2.2×10^{-4} (m^2/kg, fresh weight). Also, different time dependencies for different game animals were observed.

The Japanese government sets and operates the criteria concerning radiation and radioactive concentrations to protect people from radiation exposure. However, the regulations have affected people's lives in contaminated areas through limited access to forests, restrictions on the use of timber, wild mushrooms and plants, and wildlife. More details are described in the study by Hashimoto et al. (2022).

12.5 New Features Captured in Fukushima

The Fukushima Daiichi Nuclear Power Plant accident is the second-largest nuclear accident since the Chernobyl Nuclear Power Plant accident. There are several new features captured in the research on the Fukushima accident.

In the Chernobyl accident, few data captured the very early radiocesium dynamics in the forests (days to months after the accident). However, in the Fukushima accident, the preinstalled observation system, which had been operated for biogeochemically studying the forest before the accident, monitored the very early dynamics of the radiocesium, including deposition, trapping by the tree canopy, and migration from the canopy to the forest floor. Furthermore, the monitoring activity started early enough in many forests; thus, days to months and a few years of radiocesium in forests were well captured. These data are new and essential in areas of radioecology research, aiding the understanding of radiocesium dynamics in forests.

Another new feature is the transparency of various datasets. Various datasets about forest and forest products were released and opened by researchers and the government, and these datasets are available for researchers and the public. For example, the FFPRI and the Forestry Agency have monitored the forest radiocesium dynamics (i.e., contamination of tree organs and soils and radiation) since August 2011 and published the data annually. Fukushima Prefecture has also published similar monitoring data. The government has also released monitoring data for edible forest products, such as wild mushrooms, wild edible plants, and game meat. The airborne survey of radiation and radiocesium inventory was also demonstrated repeatedly and is available in a very easy-to-see Web interface and GIS format for specialists (e.g., https://emdb.jaea.go.jp/emdb/). These datasets have demonstrated forest contamination, as researchers have used them to study forest contamination. Furthermore, they provided the general situation of the forest contamination to the public with transparency.

12.6 Future Prediction

More than a decade of research by scientists and authorities revealed the radiocesium dynamics and radiation changes in forests. What will happen to forests in the future? Because most of the ^{134}Cs, which contributed to the rapid decrease in the radiation dose, had decayed in the last 10 years, the decrease in radiation declines and occurs with the decay of ^{137}Cs (Fig. 12.8). Thus, ^{137}Cs will continue to migrate to forests. Modeling analysis is a good tool for understanding the transfer processes and predicting future radiocesium dynamics in forests (Ota et al. 2016; Nishina et al. 2018; Thiry et al. 2018; Kurikami et al. 2019). After the Fukushima nuclear accident, modeling studies were conducted to predict radiocesium dynamics in forests, revealing that, as already demonstrated by many observations, the mineral soil

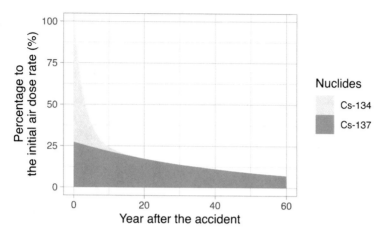

Fig. 12.8 Predicted changes in air dose rate with time based on physical decay. The ratio of ^{134}Cs and ^{137}Cs emitted at the time of the accident was assumed to be 1:1. ^{134}Cs emits radiation ~2.7 times more intense than ^{137}Cs

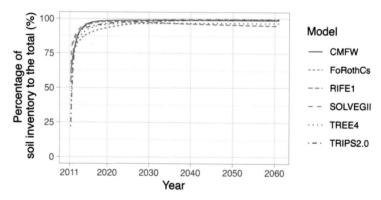

Fig. 12.9 Prediction of soil inventory percentage in coniferous forests by six models (Hashimoto et al. 2021)

compartment will be the largest radiocesium reservoir (Fig. 12.9). However, a small portion of radiocesium in the forest ecosystem circulates continuously within forests, between trees and soils, probably approximately <1% of the total initial deposition (Hashimoto et al. 2020a) (Fig. 12.10). However, the six-model intercomparison revealed that the future of radiocesium concentration in wood, which is the most important part of a tree in terms of forest products, is quite uncertain. The intercomparison demonstrates that the state-of-the-art models worked well to predict the radiocesium concentration in trees within a decade, but the uncertainty will increase with time (Fig. 12.11). These patterns were also found in a preliminary simulation for the Konara oak. In the equilibrium stage, the key driver of forest radiocesium dynamics between trees and soils is root uptake; the uncertainty and

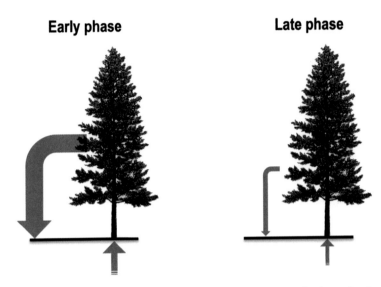

Fig. 12.10 Conceptual diagram of radiocesium circulation in forests; early phase after the fallout and later phases in the equilibrium state. In the equilibrium state, the emission from trees and the uptake from the soil are almost balanced

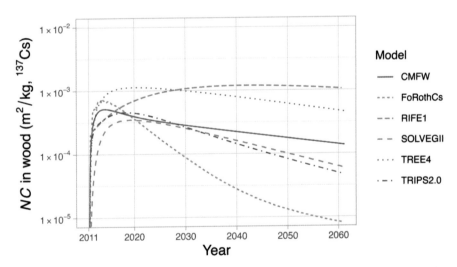

Fig. 12.11 Prediction of normalized activity concentration (*NC*) in conifer wood using six models. The activity concentration is normalized with the total initial deposition (Hashimoto et al. 2021)

variation in the model predictions are attributable to the different root uptake assumptions among the models. To reduce the future prediction uncertainty, it is essential to quantify the root uptake ratio and the biogeochemical processes of radiocesium in the soil. The radiocesium dynamics in forests will decrease with time, but will never stop. Thus, it is important to continue monitoring and under-standing the transfer process of radiocesium within forests.

Another important perspective on the future of contaminated forests is the health of forest ecosystems. Contaminated forests, particularly artificial forests, are now abandoned or undermanaged, which risks the health of forest ecosystem services. Additionally, no-go areas and less used forests have changed the balance between humans and wild animals in the affected areas. Note that these changes in forest ecosystems are not direct effects of radiation. It is crucial to monitor the forest health in affected areas and promote healthy forest functioning.

12.7 Conclusion

Immediately after the accident, the forests in Fukushima have been intensively studied. The radiocesium deposited onto the forests migrated within forests. In Fukushima forests, the dynamics were well captured by many scientific studies and monitoring works by the authorities. In the last 10 years, the total amounts of radiocesium and radiation have substantially decreased due to the decay of ^{134}Cs. In the future, the decrease will decline due to the remaining ^{137}Cs, and ^{137}Cs will circulate continuously in forests, although the amount is minimal in terms of the total reservoir in forests. Also, predictive studies using models have been conducted, which have provided useful perspectives on the dynamics of ^{137}Cs in forests. However, they still have significant uncertainty in predicting the contamination of the wood. These predictions emphasize the importance of long-term monitoring and further research on the radiocesium dynamics, such as root uptake by trees. Forest ecosystems have changed due to the change in the balance between humans and nature due to land abandonment and underuse of forests. Furthermore, it is also important to monitor and manage the health of forest ecosystems in affected areas. An even more detailed understanding of the radiocesium dynamics in forests and a reliable future prediction will help rehabilitate the environment and people's lives in the affected areas.

References

Forestry Agency (2022) Report on monitoring project of radioactivie substances in forests FY 2021. https://www.rinya.maff.go.jp/j/kaihatu/jyosen/r3_surveys_on_radioactive_cesium.html   (in Japanese)

Fukushima Prefecture Forestation Division (2021) On the situation and future prediction of radioactive substances in forests. https://www.pref.fukushima.lg.jp/uploaded/attachment/446209.pdf (in Japanese)

Gonze M-A, Calmon P, Hurtevent P, Coppin F (2021) Meta-analysis of radiocesium contamination data in Japanese cedar and cypress forests over the period 2011–2017. Sci Total Environ 750: 142311. https://doi.org/10.1016/j.scitotenv.2020.142311

Hashimoto S, Ugawa S, Nanko K, Shichi K (2012) The total amounts of radioactively contaminated materials in forests in Fukushima, Japan. Sci Rep 2:416. https://doi.org/10.1038/srep00416

Hashimoto S, Imamura N, Kaneko S et al (2020a) New predictions of ^{137}Cs dynamics in forests after the Fukushima nuclear accident. Sci Rep 10:29. https://doi.org/10.1038/s41598-019-56800-5

Hashimoto S, Imamura N, Kawanishi A et al (2020b) A dataset of ^{137}Cs activity concentration and inventory in forests contaminated by the Fukushima accident. Sci Data 7:431. https://doi.org/10.1038/s41597-020-00770-1

Hashimoto S, Komatsu M, Imamura N et al (2020c) Forest ecosystems. In: Environmental transfer of radionuclides in Japan following the accident at the Fukushima Daiichi nuclear power. IAEA, Vienna, pp 129–178. https://www.iaea.org/publications/14751/environmental-transfer-of-radio nuclides-in-japan-following-the-accident-at-thefukushima-daiichi-nuclear-power-plant

Hashimoto S, Tanaka T, Komatsu M et al (2021) Dynamics of radiocaesium within forests in Fukushima—results and analysis of a model inter-comparison. J Environ Radioact 238:106721. https://doi.org/10.1016/j.jenvrad.2021.106721

Hashimoto S, Komatsu M, Miura S (2022) Forest radioecology in Fukushima—dynamics, impact, and future. Springer, Singapore. https://doi.org/10.1007/978-981-16-9404-2

IAEA (2006) Environmental consequences of the Chernobyl accident and their remediation: 20 years of experience. IAEA, Vienna. https://www.iaea.org/publications/7382/environmen tal-consequences-of-the-chernobyl-accident-and-their-remediation-twenty-years-of-experience

Imamura N, Komatsu M, Ohashi S et al (2017) Temporal changes in the radiocesium distribution in forests over the five years after the Fukushima Daiichi nuclear power plant accident. Sci Rep 7:8179. https://doi.org/10.1038/s41598-017-08261-x

Itoh Y, Imaya A, Kobayashi M (2015) Initial radiocesium deposition on forest ecosystems surrounding the Tokyo metropolitan area due to the Fukushima Daiichi nuclear power plant accident. Hydro Res Lett 9:1–7. https://doi.org/10.3178/hrl.9.1

Kato H, Onda Y, Gomi T (2012) Interception of the Fukushima reactor accident-derived ^{137}Cs, [134]Cs and ^{131}I by coniferous forest canopies. Geophys Res Lett 39:2012GL052928. https://doi.org/10.1029/2012GL052928

Kato H, Onda Y, Gao X et al (2019a) Reconstruction of a Fukushima accident-derived radiocesium fallout map for environmental transfer studies. J Environ Radioact 210:105996. https://doi.org/10.1016/j.jenvrad.2019.105996

Kato H, Onda Y, Saidin ZH et al (2019b) Six-year monitoring study of radiocesium transfer in forest environments following the Fukushima nuclear power plant accident. J Environ Radioact 210:105817. https://doi.org/10.1016/j.jenvrad.2018.09.015

Komatsu M, Nishina K, Hashimoto S (2019) Extensive analysis of radiocesium concentrations in wild mushrooms in eastern Japan affected by the Fukushima nuclear accident: use of open accessible monitoring data. Environ Pollut 255:113236. https://doi.org/10.1016/j.envpol.2019.113236

Komatsu M, Hashimoto S, Matsuura T (2021) Effects of species and geo-information on the ^{137}Cs concentrations in edible wild mushrooms and plants collected by residents after the Fukushima nuclear accident. Sci Rep 11:22470. https://doi.org/10.1038/s41598-021-01816-z

Kurikami H, Sakuma K, Malins A et al (2019) Numerical study of transport pathways of ^{137}Cs from forests to freshwater fish living in mountain streams in Fukushima, Japan. J Environ Radioact 208:106005. https://doi.org/10.1016/j.jenvrad.2019.106005

Ministry of Land, Infrastructure, Transport and Tourism (2020) Digital national land information. https://nlftp.mlit.go.jp/ksj/ (in Japanese)

Nishina K, Hashimoto S, Imamura N et al (2018) Calibration of forest ^{137}Cs cycling model "FoRothCs" via approximate Bayesian computation based on 6-year observations from plantation forests in Fukushima. J Environ Radioact 193:82–90. https://doi.org/10.1016/j.jenvrad.2018.09.002

Ohashi S, Kuroda K, Takano T et al (2017) Temporal trends in ^{137}Cs concentrations in the bark, sapwood, heartwood, and whole wood of four tree species in Japanese forests from 2011 to 2016. J Environ Radioact 178:335–342. https://doi.org/10.1016/j.jenvrad.2017.09.008

Ota M, Nagai H, Koarashi J (2016) Modeling dynamics of ^{137}Cs in forest surface environments: application to a contaminated forest site near Fukushima and assessment of potential impacts of soil organic matter interactions. Sci Total Environ 551:590–604. https://doi.org/10.1016/j.scitotenv.2016.02.068

Shaw G (2007) Radionuclides in forest ecosystems. In: Shaw G (ed) Radioactivity in the terrestrial environment. Elsevier, Amsterdam, pp 127–155

Tagami K, Howard BJ, Uchida S (2016) The time-dependent transfer factor of radiocesium from soil to game animals in Japan after the Fukushima Dai-ichi nuclear accident. Environ Sci Technol 50:9424–9431. https://doi.org/10.1021/acs.est.6b03011

Thiry Y, Albrecht A, Tanaka T (2018) Development and assessment of a simple ecological model (TRIPS) for forests contaminated by radiocesium fallout. J Environ Radioact 190:149–159. https://doi.org/10.1016/j.jenvrad.2018.05.009

Chapter 13
Toward the Estimation of Radiocesium Activity Concentration in Trunks of Coppiced *Quercus serrata*: Leaf Availability Instead of Felling

Wataru Sakashita, Satoru Miura, Junko Nagakura, and Tsutomu Kanasashi

13.1 Introduction

Although it has been more than 10 years since the accident at Fukushima Daiichi Nuclear Power Plant (FDNPP), radiocesium (^{137}Cs) released from the plant remains in the forest. Previously, trunks of coppiced konara oak (*Quercus serrata*) growing in the Fukushima area were widely used as bed logs for the cultivation of shiitake mushroom (*Lentinula edodes*) (Miura 2016). However, the production of bed logs was stopped due to the ^{137}Cs contamination of the deciduous broad-leaved forests. The index value of 50 Bq kg^{-1} has been set for bed logs to prevent the mushrooms from exceeding the food standard limit of 100 Bq kg^{-1} (Forestry Agency 2012). The possibility exists of resuming the use of contaminated deciduous broad-leaved forests if konara oak trunks with ^{137}Cs activity concentration below this threshold can be found. Even in forests with the same degree of contamination, the extent to which konara oaks uptake ^{137}Cs from the soil is highly variable (specifically, it may differ 100 times) (Kanasashi et al. 2020). Therefore, felling surveys to measure the ^{137}Cs activity concentration in the trunk are not efficient, and developing an

W. Sakashita (✉) · S. Miura
Center for Forest Restoration and Radioecology, Forestry and Forest Products Research Institute, Tsukuba, Ibaraki, Japan
e-mail: sakashita@ffpri.affrc.go.jp; miura@ffpri.affrc.go.jp

J. Nagakura
Department of Forest Soils, Forestry and Forest Products Research Institute, Tsukuba, Ibaraki, Japan
e-mail: kurya@ffpri.affrc.go.jp

T. Kanasashi
Institute of Environmental Radioactivity, Fukushima University, Fukushima, Fukushima, Japan
e-mail: t.kanasashi@ier.fukushima-u.ac.jp

© The Author(s) 2023
T. M. Nakanishi, K. Tanoi (eds.), *Agricultural Implications of Fukushima Nuclear Accident (IV)*, https://doi.org/10.1007/978-981-19-9361-9_13

149

alternative method for estimating the ^{137}Cs activity concentration in the trunk is necessary.

To achieve this goal, our research group focused on the use of the "current-year branch" in the dormant stage as an indicator of root uptake of ^{137}Cs from the soil to tree (Kanasashi et al. 2020). The movement of ^{137}Cs in the tree is considered to be stable in the dormant stage. In addition, Kobayashi et al. (2019) confirmed that ^{137}Cs activity concentration in the trunk is positively correlated with that of current-year branch during this dormant stage. Based on the above findings, we are developing a method using current-year branch in the dormant stage to estimate the ^{137}Cs activity concentration in the trunk without felling.

Here, we assume that the survey of the current-year branches of konara oaks generally takes place during the dormant stage, when tree growth generally stopped. This limited period was one of the issues in developing our alternative method, given that the felling of coppiced konara oaks is generally performed in autumn and winter, and judging the degree of contamination is necessary before felling takes place. Accordingly, we assessed the seasonal stability of ^{137}Cs in the current-year branches of coppiced konara oaks (Sakashita et al. 2021). We found that the period for surveying the current-year branches can be approximately doubled and that current-year branches before the felling season can be used for estimating the ^{137}Cs activity concentration in the trunk in Miyakoji, Tamura, Fukushima Prefecture. This new finding is expected to be useful in related investigations that require a survey of ^{137}Cs activity concentration in the trunk. However, it is not always possible for those in charge of the logging operations to correctly identify the current-year branches. Therefore, instead of current-year branches, another index is required as a proxy measure for the estimation of the ^{137}Cs activity concentration in the trunk.

Taking into account these recent developments, our research group focused on the potential of using leaves as data sources. Sakashita et al. (2021) reported not only the seasonal stability of current-year branch ^{137}Cs activity concentration but also the relation between ^{137}Cs activity concentration in the leaves and the corresponding values in the current-year branches. This relation is expected to be stable from the end of the flushing stage to prior to the defoliation stage. From this result, we newly hypothesized that the ^{137}Cs activity concentration in the leaves at this phenological stage was as stable as that of the current-year branches; thus, the leaves could be used to estimate the ^{137}Cs activity concentration in the trunk. If this hypothesis is true, estimating the ^{137}Cs activity concentration in the trunk more easily will be possible, because the leaves can be sampled more easily than current-year branches. In addition, the damage to the coppiced oaks from leaf sampling is probably small compared to the damage from sampling current-year branches. Because of the recent interest in estimating the ^{137}Cs activity concentration in the trunk using current-year branches in the dormant stage, we decided to verify whether ^{137}Cs activity concentration of current-year branches in the dormant stage can, in fact, be estimated from sampling the leaves.

Our study took place in Miyakoji, Tamura, Fukushima Prefecture. To test our hypothesis, we determined the period in which the ^{137}Cs activity concentration in the leaves of konara oaks in the study area is stable. Second, we examined the relation between the ^{137}Cs activity concentration in the current-year branches in the dormant stage and the corresponding values in the leaves obtained at the phenologically stable stage. Finally, we verified whether the ^{137}Cs activity concentration in current-year branches in the dormant stage can be actually estimated from the leaves.

13.2   Materials and Methods

13.2.1   Data

We used the ^{137}Cs activity concentration data (decay corrected to June 1, 2018) of the current-year branches and leaves (Fig. 13.1a, b) in coppiced konara oak (*Q. serrata*) reported by Sakashita et al. (2021). Here 1-year branches after the dormant stage were defined as the "current-year branches," and "dormant stage" was defined as from November to April of the following year based on the analysis of the current-year branch ^{137}Cs activity concentration (Sakashita et al. 2021). The data from 2018 to 2020 (sampling intervals: from 10 to 98 days; intensive sampling: from April to July in 2020) were obtained from six study sites in Miyakoji, Tamura, Fukushima Prefecture (Fig. 13.1c). Konara oaks at these sites regenerated after the

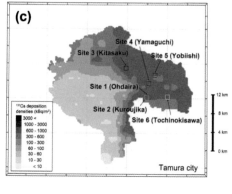

Fig. 13.1 (**a**) Sample photo before dividing. (**b**) Sample photo after dividing into three parts (leaves, current-year branches, and previous-year branches). (**c**) Map showing six study sites in Miyakoji, Tamura, Fukushima Prefecture. Darker shades of gray indicate higher ^{137}Cs deposition densities (decay corrected to December 28, 2012) reported by Mext (Ministry of Education, Culture, Sports, Science and Technology, Japan) (2013). Local place names are given in parentheses. [This map was modified from Sakashita et al. (2021)]

FDNPP accident and had been coppiced (2011–2015). The data from four study sites (Sites 1–4) were used to evaluate the seasonal stability of leaf ^{137}Cs activity concentration and the relation between the current-year branch in the dormant stage and leaf ^{137}Cs activity concentration. Based on these relations, data from the remaining two study sites (Sites 5–6) were used to verify whether the current-year branch ^{137}Cs activity concentration in the dormant stage could actually be estimated from the ^{137}Cs activity concentration in leaves.

13.2.2 Analysis

To evaluate seasonal stability of ^{137}Cs in leaves at Sites 1–4, the values of [137]C activity concentration in leaves were normalized to zero mean and one standard deviation. The means and standard deviations were calculated using the data from May 28 to September 15, 2019, which was the only period in which samplings were performed at the same time at all four study sites (Sites 1–4). Then, we assessed the seasonal stability of ^{137}Cs activity concentration in leaves by merging 3-year observation data on a horizontal axis of 365 days per year. This horizontal axis begins on May 1, because the leaves of konara oaks in Miyakoji generally start to open (flushing) beginning in May. As an additional note, we assumed in this analysis that no interannual variation was found in the ^{137}Cs activity concentration of the leaves.

At four study sites (Sites 1–4), we assessed the relation between ^{137}Cs activity concentration in the current-year branch (collected in the dormant stage) and corresponding values in the leaves; we applied robust regression, in which the effects of outliers can be reduced (MathWorks 2021). The regression coefficient (±standard error) was used as the coefficient for estimating the ^{137}Cs activity concentration in the current-year branches from the leaves. Using the data from Sites 5–6, we also verified whether ^{137}Cs activity concentration in the current-year branches can actually be estimated from the ^{137}Cs activity concentration of the leaves (collected in July, August, and October) based on this robust regression.

13.3 Results and Discussion

13.3.1 Seasonal Stability of Leaf [137]Cs Activity Concentration

At four study sites (Sites 1–4), the normalized ^{137}Cs activity concentration (normalized to zero mean and one standard deviation) in leaves indicated that the concentration is highest in early May and then decreases rapidly until early July (Fig. 13.2). This period from May to June (or early July) corresponds to the flushing stage of konara oak in the study area; a similar trend can be seen in the seasonal variation of ^{137}Cs activity concentration in current-year branches (Sakashita et al. 2021). Given

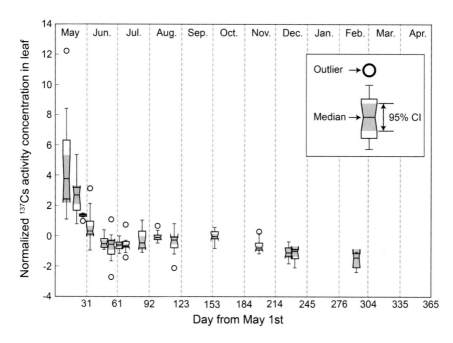

Fig. 13.2 Seasonal variation of normalized leaf ^{137}Cs activity concentration at four study sites (Sites 1–4). [137]Cs activity concentrations are normalized to zero mean and one standard deviation from May 28 to September 15, 2019, and decay was corrected to June 1, 2018. Lines inside the boxes indicate the medians, and the upper and lower boundaries of the boxes represent the 25th and 75th percentiles, respectively. Whiskers extend to show the data distribution within 1.5 × interquartile range. Open circles indicate the outliers, and gray shadings indicate the 95% confidence intervals (CI) of the medians

that the seasonal movement of ^{137}Cs in the tree is considered to be most active in this stage, using the leaves in this flushing stage is probably not feasible for estimating the current-year branch ^{137}Cs activity concentration (in the dormant stage).

From late July, after the flushing stage, we found that the normalized ^{137}Cs activity concentration in leaves was stable until the beginning of October; the median 95% confidence intervals (CI) overlapped during this period (Fig. 13.2). In addition, plentiful leaves are found in the canopy during this period (Fig. 13.3a–d); thus, obtaining enough leaf samples to measure the ^{137}Cs activity concentration is easy. Accordingly, the leaves sampled during this period are considered suitable for estimating the ^{137}Cs activity concentration of the current-year branches.

The normalized ^{137}Cs activity concentrations in leaves from November to February were significantly lower (at 5% significant level) than the corresponding values from late July to early October (Fig. 13.2). This is thought to be due to the translocations of cesium from the leaves to other tree organs before the dormant stage (e.g., Yoshihara et al. 2019; Kenzo et al. 2020). The ^{137}Cs activity concentration was also stable from November to February; however, in the study area

Fig. 13.3 Photos of konara oaks (*Q. serrata*) at Miyakoji (Tamura, Fukushima Prefecture) on July 25, 2018 (**a**), August 8, 2019 (**b**), September 20, 2016 (**c**), October 3, 2018 (**d**), November 15, 2018 (**e**), and December 20, 2019 (**f**)

(Miyakoji, Tamura), most of the leaves in the canopy generally fall after November, after which it is not feasible to obtain a large enough sample of leaves (Fig. 13.3e, f).

13.3.2 ^{137}Cs Activity Concentration in Current-Year Branches, Estimated Using Sampled Leaves

At Sites 1–4, where we previously evaluated the seasonal stability of ^{137}Cs activity concentration in the leaves of konara oak, we found that the average leaf ^{137}Cs activity concentration from late July to October was significantly correlated with that of current-year branches (Fig. 13.4; $R^2 = 0.93$, $p < 0.001$). The regression coefficient ± standard error was 0.54 ± 0.03, and the intercept of the regression line was 8.5; 95% CI of the intercept ranged from −57 to 74. In order to simplify the estimation method, the intercept was set to zero since no significant difference from zero was found.

At Sites 5–6, we examined whether ^{137}Cs activity concentration in the current-year branches can be estimated solely by multiplying the above regression coefficient (0.54 ± 0.03) by the leaf ^{137}Cs activity concentration during late July–October. Our comparisons between estimated and measured ^{137}Cs activity concentration in the current-year branches indicated that 99% CI of the robust regression lines overlapped one-to-one correspondence lines on July 25, August 25, and October 3, 2018 (Fig. 13.5). Although the verification data set is relatively small, the leaves obtained on these dates could actually be used to estimate ^{137}Cs activity concentration in the current-year branches (from the dormant stage).

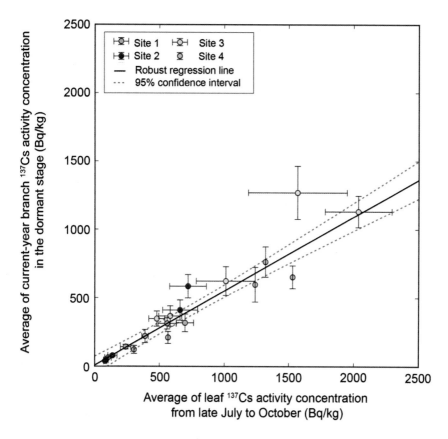

Fig. 13.4 Relation between average of ^{137}Cs activity concentration in leaves from late July to October and average of current-year branch ^{137}Cs activity concentration in the dormant stage at Sites 1–4 (cyan circle: Site 1; red circle: Site 2; yellow circle: Site 3; gray circle: Site 4). Error bars represent the standard deviations. No error bar is found on the horizontal axis of Site 4, because the leaves were collected only once from July to October. Black solid line indicates the robust regression line, and black dotted lines show the 95% CI of the regression line

Considering the seasonal stability of leaf ^{137}Cs activity concentration values in the above three periods (May–early July [flushing stage], late July–October [stable stage], and November–February [defoliation and dormant stage]) and the ease of sampling, only one period—from late July to the beginning of October—is thought to be suitable to estimate the ^{137}Cs activity concentrations of current-year branches in the dormant stage in Miyakoji. This period generally corresponds to the time after the flushing stage and before defoliation. Therefore, if these findings are applied to the areas with local climates different from that in Miyakoji, we suggested that the leaves should be obtained from the time when the leaves almost stop growing to the time before the leaves become senescent (before the leaves begin to change colors).

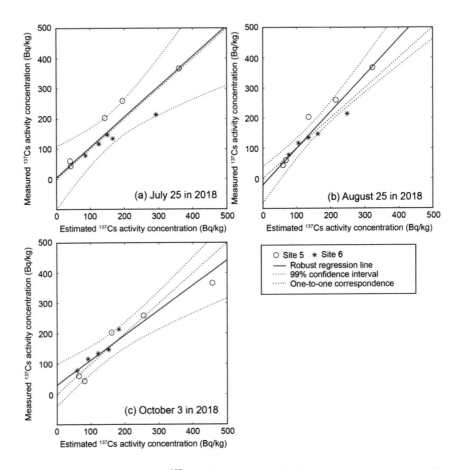

Fig. 13.5 Relation of estimated ^{137}Cs activity concentration in current-year branches and the measured values during the dormant stage at Sites 5 and 6 (black circles: Site 5; black asterisks: Site 6). Here, the ^{137}Cs activity concentration of current-year branches (from the dormant stage) was estimated using leaves sampled in 2018, on July 25 (**a**), August 25 (**b**), and October 3 (**c**). Black dotted lines indicate one-to-one correspondence. Red solid lines are robust regression lines, and red dotted lines show 99% CI of the regression lines

13.4 Conclusions

In this study, we investigated the seasonal stability of ^{137}Cs activity concentration in the leaves of coppiced konara oaks, as well as leaf availability, to estimate the ^{137}Cs activity concentration of current-year branches taken in the dormant stage. The seasonal variation in the ^{137}Cs activity concentration in leaves and the method for estimating the ^{137}Cs activity concentration of current-year branches are summarized in Table 13.1. Our results indicated three distinct stages in seasonal variation of ^{137}Cs activity concentration in leaves: (1) flushing stage, (2) stable stage, and (3) defoliation and dormant stage. In the flushing stage (from May to early July),

Table 13.1 Summary of seasonal variations in leaf ^{137}Cs activity concentration in coppiced konara oaks and availability of leaves for estimating current-year branch ^{137}Cs activity concentration in the dormant stage. Seasonal variations of current-year branch ^{137}Cs activity concentration and the potential for estimating these values for the dormant stage are also described

Tree compartment	Period[a]	Seasonal variation	Availability for estimating current-year branch ^{137}Cs activity concentration in the dormant stage
Leaf (this study)	May–early July (flushing stage)	Rapid decrease	Not available due to the large seasonal variation of ^{137}Cs in the tree
	Late July–October (stable stage)	Stable	Available by multiplying 0.54 ± 0.03
	November–February (defoliation and dormant stage)	Stable	Not practical due to the difficulty of sampling (the number of leaves in canopy is relatively very less, with few leaves remaining after the month of November)
Current-year branch Sakashita et al. (2021)	May–June (flushing stage)	Rapid decrease	Not available due to the large seasonal variation of ^{137}C in the tree
	July–October (gradual growing stage)	Gradual decrease	Available by multiplying 0.75 ± 0.10[b]
	November–April (dormant stage)	Stable	–

[a]In the areas where the climate is different from that of Miyakoji, Tamura, Fukushima Prefecture, there may be differences in the above periods
[b]There is a limitation to the July correction (<200 Bq kg^{-1})

^{137}Cs activity concentration decreased rapidly and was not stable, suggesting that it is not suitable for estimating ^{137}Cs activity concentration in current-year branches. In the stable stage (from late July to October), ^{137}Cs activity concentration values in leaves were stable. Also, we confirmed that ^{137}Cs activity concentration of current-year branches (in the dormant stage) can be estimated by multiplying ^{137}Cs activity concentration in leaves obtained in the stable stage by the coefficient 0.54 ± 0.03. We showed that estimating ^{137}Cs activity concentration in current-year branches is possible by using the corresponding values in sampled leaves as a proxy measure. In the defoliation and dormant stage (from November to February), although ^{137}Cs activity concentration in leaves at this stage was significantly lower than during stable stage, the activity concentration was also stable during this stage. However, we concluded that it is not practical to use leaves in this season to estimate current-year branch ^{137}Cs activity concentration, given that few leaves are remaining in the canopy at this stage.

In summary, we report that, in the case of konara oaks in Miyakoji, Tamura, Fukushima Prefecture, the ^{137}Cs activity concentration of current-year branches (in dormant stage) can be estimated by sampling leaves in the phenologically stable stage (from late July to October). This finding suggests that material from leaves can also be used as a method for estimating ^{137}Cs activity concentration of the trunk without felling the tree. Because leaves are more easily to sample than the current-year branches, this finding exhibits important and practical implications in the search for the konara oaks that can be used as bed logs for shiitake mushroom cultivation in the contaminated area of Fukushima.

Acknowledgements The authors are grateful to the Fukushima Chuo Forestry Co-operative for their cooperation. We would also like to thank the staff of the Forestry and Forest Products Research Institute (FFPRI) for their assistance. This work was supported by research grants from the Bio-oriented Technology Research Advancement Institution, NARO, Japan (a research program for the development of innovative technology; 28028C) and from FFPRI (#201901).

References

Forestry Agency (2012) Establishment of index values for mushroom logs and cultivars, and firewood and charcoal for cooking and heating. In: Sectoral information. https://www.rinya.maff.go.jp/j/tokuyou/shihyouti-index.html

Kanasashi T, Miura S, Hirai K, Nagakura J, Itô H (2020) Relationship between the activity concentration of ^{137}Cs in the growing shoots of *Quercus serrata* and soil ^{137}Cs, exchangeable cations, and pH in Fukushima Japan. J Environ Radioact 220:106276

Kenzo T, Saito S, Miura S, Kajimoto T, Kobayashi NI, Tanoi K (2020) Seasonal changes in radiocesium and potassium concentrations in current-year shoots of saplings of three tree species in Fukushima. Jpn J Environ Radioact 223:106409

Kobayashi NI, Ito R, Masumori M (2019) Radiocesium contamination in forests and the current situation of growing oak trees for mushroom logs. In: Nakanishi TM, O'Brien M, Tanoi K (eds) Agricultural implications of the Fukushima nuclear accident (III) after 7 years. Springer, Singapore, pp 107–122

MathWorks (2021). https://www.mathworks.com/help/stats/robust-regression-reduceoutlier-effects.html

Mext (Ministry of Education, Culture, Sports, Science and Technology, Japan) (2013) Results of deposition of radioactive cesium of the sixth airborne monitoring survey and airborne monitoring survey outside 80 km from the Fukushima Daiichi NPP

Miura S (2016) The effects of radioactive contamination on the forestry industry and commercial mushroom-log production in Fukushima, Japan. In: Nakanishi TM, Tanoi K (eds) Agricultural implications of the Fukushima nuclear accident the first three years. Springer, Tokyo, pp 145–160

Sakashita W, Miura S, Nagakura J, Kanasashi T, Shinomiya Y (2021) Seasonal stability of ^{137}Cs in coppiced *Quercus serrata* current-year branches: toward the estimation of trunk ^{137}Cs activity concentrations without felling. Ecol Indic 133:108361

Yoshihara T, Yoschenko V, Watanabe K, Keitoku K (2019) A through year behavior of ^{137}Cs in a Japanese flowering cherry tree in relation of that of potassium. J Environ Radioact 202:32–40

Chapter 14
Decomposition of Organic Matters in a Forest Floor Enhanced Downward Migration of Radioactive Cs After the Accident of the FDNPP

Taku Nishimura, Shoichiro Hamamoto, Takuhei Yamasaki, and Takahiro Tatsuno

14.1 Introduction

Radioactive materials emitted by the Fukushima Daiichi Nuclear Power Plant (FDNPP) accident in March 2011 are still causing environmental issues in northeastern Japan. The main element of concern is cesium (Cs), which was not often studied in soil science. Cesium behaves differently from Na, Ca, and other metal cations commonly noted in soil physics. The range of Cs concentrations in the environment is considerably lower than that of most cations observed in previous studies on soil solute transport. For example, one million becquerels per kg of soil (Bq kg^{-1} hereafter) is equivalent to 0.3 ppb in mass. The distribution coefficient of Cs in soil under the ppb concentration range is far greater than that under the ppm order (Comans et al. 1991). Notably, the distribution coefficient of Cs is concentration dependent.

The half-life of ^{137}Cs is relatively longer than that of other radioactive materials emitted from the accident; therefore, pollution from ^{137}Cs and countermeasures must occur over a long period. However, because of cost factors, decontamination of radioactive Cs in forests was limited to the perimeter of the forests, and the fate of ^{137}Cs deposited in forests in Fukushima is an ongoing issue for pollution management.

T. Nishimura (✉) · S. Hamamoto · T. Yamasaki
Laboratory of Soil Physics and Soil Hydrology, Department of Biological and Environmental Engineering, Graduate School of Agricultural and Life Sciences, The University of Tokyo, Tokyo, Japan
e-mail: atakun@g.ecc.u-tokyo.ac.jp

T. Tatsuno
Institute of Environmental Radioactivity, Fukushima University, Fukushima City, Fukushima, Japan

© The Author(s) 2023
T. M. Nakanishi, K. Tanoi (eds.), *Agricultural Implications of Fukushima Nuclear Accident (IV)*, https://doi.org/10.1007/978-981-19-9361-9_14

161

Cs is a monovalent cation with a large ionic radius (0.169 nm), which results in more significant dehydration than other monovalent cations. Dehydrated Cs cations are often found in the ditrigonal siloxane cavity at the surface of clay plates and frayed edge sites between clay layers. This essentially causes irreversible adsorption of Cs onto clay (Dalvaux et al. 2000). In addition to the structural characteristics of clay minerals, dehydrated Cs exhibits thermodynamic ion-exchange characteristics quite different from those of sodium and divalent cations. As a result Cs prefers adsorption to the solid phase (Appelo and Postma 2005).

A greater distribution coefficient in the solid phase causes retardation of Cs transport in soils. After the FDNPP accident, deposited radioactive Cs was selectively detected in near-surface soil, and its downward migration was slow (Takahashi et al. 2018). After the accident, the government focused on the delay in Cs transport and assumed that radioactive Cs mainly stayed near the surface. Large-scale surface soil scraping was performed based on this assumption. However, the preference for adsorption to solids suggests the possibility of colloid-facilitated transport of radioactive Cs in soil.

This study focused on organic colloids produced by the degradation of surface litter that may enhance the migration of Cs in soils. Soil samples were collected from the surface litter layer on the forest floor, and bare land adjacent to it, and the distribution of radioactive Cs was investigated.

14.2   Study Site and Methods

The detailed geomorphology of the Fukushima slope studied is shown in Fig. 14.1. The slope was located approximately 30 km northwest of FDNPP, and had a deposition of radioactive Cs that was delivered by the wind from the southeast on the day of the accident. The region was covered with weathered granite soil.

In May 2013, we collected undisturbed soil column samples (5 cm in diameter and 30 cm in length) using a liner soil sampler (DIK-110C, Daiki Inc., Saitama, Japan). Similar sampling was conducted in the upslope area in 2014.

The samples were taken to the laboratory. Layers of the soil column were photographed, and the soil was cut at predetermined intervals, and then, the radioactive Cs concentration, soil organic carbon content, and soil nitrogen content were analyzed for each depth. For the 20 cm thick surface layer, the soil was cut at 2 cm intervals, with larger intervals for the deeper layer. The sliced samples were oven-dried and ground with a mortar and pestle, and stones and coarse organic matter were removed. The sample was then placed in a 20 mL vial, and the radioactive Cs content was measured using an NaI spectrophotometer (Wizard 2480, Perkin Elmer Inc.). After radioactivity measurement, a sample was taken from the vial, and the organic carbon and nitrogen contents were measured using a CN coder (CN-60, Sumika Inc.).

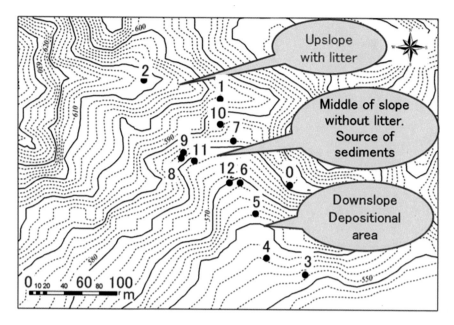

Fig. 14.1 Geomorphological features of the study site

14.3 Results and Discussion

14.3.1 Surface Conditions and Soil Layers

Photo 14.1 shows the sample obtained in May 2013. Weathered granite soil dominated the area. In general, yellow-weathered granite soil was observed in low organic content subsoil and outcrops of cliffs. The dark color of the soil suggests the presence of organic matter.

In Fig. 14.1, Sites 1, 2, and 8 are soils from the upslope forest floor with a surface litter layer. Neither trace of water erosion nor sediment deposition was observed in the upslope area. Sites 1 and 2 showed a dark color near the surface, which became lighter in the deeper layer. The soil sample from Site 8 showed dark colors from the surface to a depth of 30 cm. Mid-slope areas (Sites 7, 9, 10, and 11) were under complex erosion and sedimentation processes. Some sites experienced erosion, whereas others revealed sediment deposition from the upslope. The mid-slope area was covered with weeds and young tree bush. There was a partial organic matter on the soil surface, but no significant litter layer, such as that observed at the upslope. Site 9 had bird feathers found at a depth of 25 cm, suggesting that most of the soil had been deposited recently. Thus, we omitted soil from Site 9 from further analysis. Soil samples from the mid-slope area showed a dark color near the surface, and yellow soil was strongly affected by the parent material from 13 cm (Sites 10 and 11) to 20 cm (Site 7) in depth. Sediments from the upslope were deposited at the foot of

Photo 14.1 Photograph of strata of soil cores sampled in 2013. The number in the photograph denotes the sampling site in Fig. 14.1

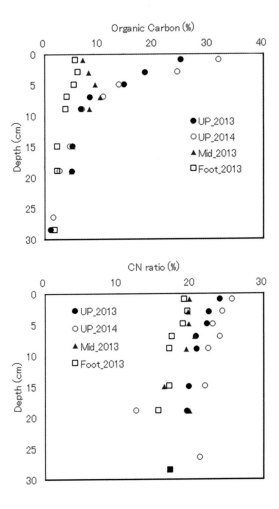

the slope (Sites 3, 4, 5, and 6). Site 3 had black soil in the near-surface layer and the layer deeper than 20 cm. This deeper dark layer suggests it was the former soil surface, and organic matter had accumulated at that time. The dark layer near the surface at Site 3 could be recently deposited sediments from upstream. Site 4 showed different textured loose soils in the 0–5 cm and 5–15 cm layers. This reflects recent sediments from the upslope and may be due to the loose soil structure in the 10 cm deep layer at Site 6. Site 5 showed an accumulation of organic matter and a dark color for the entire layer, from the surface to a depth of 30 cm. Site 5 may have been one of the recessed depositional areas in this slope.

14.3.2 Organic Carbon and Carbon/Nitrogen Ratio

The average organic carbon content and carbon/nitrogen (CN) ratio distributions for each area are shown in Fig. 14.2. The upstream values, UP_2013 and UP_2014, were averages of Sites 1, 2, and 8 in 2013 and 2014, respectively, while midstream (Mid_2013) and downstream (Foot_2013) showed averages for mid-slope areas at Sites 7, 10, and 11, and downstream at 3, 4, 5, and 6, respectively. The upstream area showed a high organic carbon content at 0–8 cm in depth, where a fibrous structure is observed in Photo 14.1. Organic carbon content decreased with increasing depth. The CN ratio of soil from the upslope area decreased slightly with increasing depth. The organic carbon content distribution of the upslope area was similar to that of nonvolcanic ash soil reported for a forest floor in western Japan (Kawahara 1970). No trace of water erosion was observed; therefore, the decomposition of organic materials in surface litter likely supplied secondary organic compounds to the soil layer. Organic compounds migrated to deeper layers with further decay (Kawahara 1970), and this further decay caused a slight decrease in the CN ratio. Secondary organic compounds could be what is generally referred to as dissolved organic carbon in the soil.

The soil sampled at the mid-slope did not show high organic carbon content near the surface, as observed in the upslope area. At approximately 8 cm in depth and deeper, the mid-slope soil showed similar organic carbon levels and CN ratios to the one at the upslope, and both decreased in the deeper layers. The lower organic carbon content for the surface at 0–8 cm layer suggested that removing surface soil by erosion affected the 8 cm thick surface layer, and fresh organic matter with a higher CN ratio was essentially removed. This could have caused low organic carbon in the soil near the surface.

Downstream of the slope, Sites 3, 4, 5, and 6 showed relatively narrow ranges of organic carbon content and CN ratio distributions, possibly due to the deposition of sediments from the upslope.

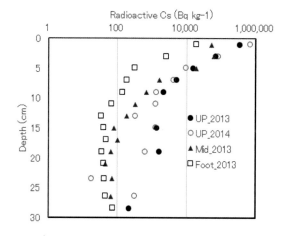

Fig. 14.2 Depth distribution of soil organic carbon and carbon-to-nitrogen (CN) ratio on slopes. UP, Mid, and Foot indicate the average of upstream (1, 2, 8), midstream (7, 10, 11), and downstream (3, 4, 5, 6) samples. The number in the legend indicates the sampling year

14.3.3 Radioactive Cs Distributions

Radioactive Cs is rare in nature. The measurement of radiation by an NaI scintillation spectrometer does not give us as precise a measurement as that of a Ge-semiconductor detector. However, dozens or a hundred Bq kg^{-1} of radioactive Cs detected in the environment using the NaI spectrometer can be traces of a particular event. Soils at the mid-slope and downstream of the slope showed low radioactive Cs in layers deeper than 15 cm, which may indicate the initial soil Cs prior to the FDNPP accident.

The radioactive Cs distribution in the three areas of the slope is shown in Fig. 14.3. Soil samples from the upslope showed high radioactive Cs content for a near-surface layer, with greater than 10,000 Bq kg^{-1} in layers above 5 cm in depth, and greater than 1000 Bq kg^{-1} at depths greater than 10 cm. The mid-slope area showed a similar radioactive Cs content distribution as the upslope area for the shallow soil layer (0–10 cm in depth). However, the mid-slope area had a further decrease in radioactive Cs content in the deeper layer, whereas the upslope maintained a greater amount up to 20 cm in depth. Downstream of the slope, a relatively large radioactive Cs content distribution was observed in the shallow layer; however, it was less than 100 Bq kg^{-1} at depths below 10 cm. Photo 14.1 suggests that downstream of the slope had a loose soil structure at a 10 cm thick surface layer, which could indicate that depositional sediments were the main constituents. The sediments from the upslope area had radioactive Cs, although no further downward migration of the radioactive Cs was observed in the downstream area.

Erosion and sedimentation occurred in the mid-slope area. Photo 14.1 suggests that the 5–10 cm thick surface soils from the mid-slope area were recently deposited sediments, because the structure appeared to be loose. As shown in Fig. 14.3, this layer had a higher radioactive Cs content. Below the sedimentation layer, at a depth greater than 10 cm at Site 10 in Photo 14.1, the radioactive Cs content was distinctly

Fig. 14.3 Depth distribution of radioactive Cs on slopes. UP, Mid, and Foot indicate the average of upstream (1, 2, 8), midstream (7, 10, 11), and downstream (3, 4, 5, 6) samples. The number in the legend indicates the sampling year

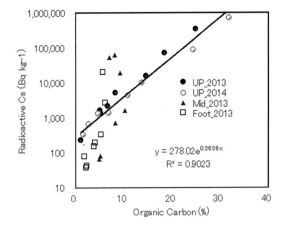

$y = 278.02e^{0.2628x}$

$R^2 = 0.9023$

less than that in the upslope area. Additional specific processes must have enhanced the downward migration of radioactive Cs in the soil in the upslope area.

Dissolved Cs move more slowly than water in the soil, and the extent of the delay can be evaluated using the retardation factor (R) as follows:

$$R = 1 + \frac{\rho_d}{\theta} K_d,$$

where K_d is the distribution coefficient (L kg^{-1}), ρ_d is the dry bulk density, and θ (m^3 m^{-3}) is the volumetric water content.

Assuming the distribution coefficient of Cs was 1000 Bq kg^{-1} as a conservative number (Ishikawa et al. 2007), the dry bulk density of the soil (1.3 g cm^{-3}) and porosity (0.5) produce a retardation factor of 2600 (Appelo and Postma 2005). This suggests that even in a water-saturated condition, radioactive Cs move at a velocity 1/2600 times slower than that of water in the soil. If the soil is unsaturated, the migration of Cs is considerably slower.

Assuming that the annual precipitation of the region is 1370 mm year^{-1} (Meteorological Agency of Japan) and evapotranspiration is 400 mm year^{-1} (Kohatake 1989), the average downward water flux density would be 970 mm year^{-1}. The porosity of the soil is 0.5; therefore, the average water flow velocity within the soil would be calculated at 970/0.5 = 1940 mm year^{-1}. Using the retardation factor, the estimated migration velocity of Cs is 1940/2600 = 0.75 mm year^{-1}.

The estimated migration of the dissolved radioactive Cs was 1.5 mm for 2 years, which is less than 1/100 of that observed for the upslope area, with radioactive Cs content near 1000 Bq kg^{-1} at depths of 20 cm. Therefore, the discussion of dissolved Cs alone cannot explain the large migration of radioactive Cs after the FDNPP accident.

14.3.4 Effects of Surface Litter on the Migration of Radioactive Cs

It was interesting that the soil near the surface above 10 cm in depth at the upslope and mid-slope areas showed high and similar radioactive Cs content distribution, while the upslope area alone still had a high content at those deeper than 10 cm. The organic matter near the surface layer could be a reason for the difference in radioactive Cs distribution between upslope and mid-slope soils.

Figure 14.4 shows the relationship between the radioactive Cs and organic carbon contents of the soil samples. For the upslope area covered by litter with no distinct trace of water erosion, the radioactive Cs content was correlated with soil organic carbon content. The correlation was valid for both soil samples collected in 2013 and 2014, with a high coefficient of determination. Soils from the mid-slope and downstream areas did not show a significant correlation with the organic carbon content (regression is not shown in Fig. 14.4). Complex erosion and deposition

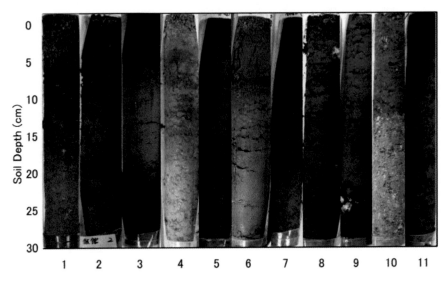

Fig. 14.4 Correlation between organic carbon content and radioactive Cs of soil samples. UP, Mid, and Foot indicate the average of the upstream (1, 2, 8), midstream (7, 10, 11), and downstream (3, 4, 5, 6) samples. The number in the legend indicates the sampling year

processes may have contributed to this result. The mid-slope and downstream regions showed some depositional sediment near the surface. This could disturb the distribution of radioactive Cs in the soil and, thus, their relationship with the organic carbon content.

As mentioned previously, the soil organic carbon and CN ratio distributions of the upslope area were similar to those observed on a forest floor in western Japan under similar conditions of surface litter, nonvolcanic ash soil, and a moist climate. Dissolved and colloidal organic matter can be produced and supplied to the soil by surface organic matter decomposition, which could affect radioactive Cs distribution at the study site. The surface litter had a high amount of radioactive Cs deposited during the FDNPP incident. Decomposition of polluted litter could release radioactive Cs from organic matter, which then migrated into deeper layers in dissolved or colloidal form. In addition, dissolved organic matter produced through surface litter decomposition may also affect the migration of Cs in soil.

Recently, Tatsuno et al. (2022) conducted a column experiment that had different orders of Cs and organic matter load to investigate the role of dissolved organic matter in the migration of radioactive Cs with a concentration found in the natural environment. Those authors compared the migration of Cs in a Cs-dissolved organic matter mixed solution and a Cs solution applied after the addition of a certain amount of dissolved organic matter. They concluded that dissolved organic matter could enhance the migration of radioactive Cs in the soil. This was interpreted as a mask of strong adsorption sites such as the FES by dissolved organic materials, and colloids facilitated the transport of Cs by dissolved organic matter, which enhanced Cs migration. Such mechanisms are expected to play a role at our study site. Concerning

Tatsuno et al. (2022), it can be considered that two processes contributed to enhancing the migration of Cs in our study. The long-term supply of dissolved organic matter by the decomposition of surface organic residue and litter that started before the FDNPP accident could have masked the strong adsorption sites of clay minerals such as FES. This could have enhanced the migration of deposited radioactive Cs near the surface soil. After the accident, the decomposition of newly deposited organic materials, which contained radioactive Cs emitted by the accident produced dissolved and colloidal radioactive Cs through the decomposition process. The dissolved Cs showed more downward migration than that reported in previous studies (Yamaguchi et al. 2012), since adsorption sites of the soil near the surface were masked due to the long-term supply of dissolved organic matter. As a result, the upslope and mid-slope areas showed similar radioactive Cs distributions below the litter layer, at approximately 4–10 cm in depth. Decomposition of newly deposited organic matter could also produce colloidal Cs. In addition to Tatsuno et al. (2022), previous studies have reported a decrease in the retardation factor of radioactive elements (Pu, Am) by the addition of colloidal particles into the solution (i.e., Contardi et al. 2001), and an increase in the migration of Cs through the soil (Turner et al. 2006). Organic colloids facilitated Cs transport and were expected to cause the further downward migration of radioactive Cs in the upslope area. As mentioned previously, the concentration ranges of common cations and radioactive Cs are different. One million Bq kg^{-1} of radioactive Cs is equivalent to 0.3 ppb in mass. When organic colloids facilitated Cs migration, a slightly undetectable amount of organic colloids was effective in altering the radioactive Cs distribution in the soil. Even organic carbon contents of deeper layers showed no distinct difference between the upslope and mid-slope area colloids produced by the decomposition of litter, which was expected to contribute to the further migration of Cs into the deeper layer.

14.4 Conclusion

Cs preferred to adsorb to solid phases such as soil particles, and radioactive Cs emitted by the FDNPP were thought to have stayed near the surface soils. However, we observed a greater than 20 cm downward migration of radioactive Cs during the 2 years after the accident at our study site. The soil radioactive Cs content was highly correlated with the soil organic carbon content. Therefore, the dissolved and/or colloidal organic matter produced by the decomposition of the forest floor surface litter played a role in the enhanced migration of radioactive Cs. The mechanisms of the enhanced transport require further investigation.

Acknowledgements The authors would like to thank the Japan Society of Promotion Science for supporting this study (19H00958, 15H02467). To conduct field surveys for this study, we received help from the NPO Resurrection of Fukushima and former staff and students of the Laboratory of Soil Physics and Soil Hydrology, The University of Tokyo. Radioactivity measurements were conducted with support from the Isotope Facility for Agricultural Education and Research, The University of Tokyo.

References

Appelo CAJ, Postma D (2005) Chapter 6: Ion exchange. In: Geochemistry, groundwater and pollution, 2nd edn. Balkema, London

Comans RNJ, Haller M, de Preter P (1991) Sorption of cesium on Illite: non-equilibrium behavior and reversibility. Geochim Cosmochim Acta 55:433440

Contardi JS, Turner DR, Ahn TM (2001) Modeling colloid transport for performance assessment. J Contam Hydrol 47:323–333

Dalvaux B, Kruyts N, Cremer A (2000) Rhizospheric mobilization of radiocesium in soils. Environ Sci Technol 34:1489–1493

Ishikawa N, Uchida S, Tagami K (2007) Effects of clay minerals on radiocesium sorption behavior onto paddy field soil. Radioisotopes 56:519528

Kawahara T (1970) A study on the amount of mineral-nitrogen in forest soil (II) the relationships between mineralized nitrogen and total nitrogen, C/N and mineralized carbon. J Jpn For Soc 52(3):71–19

Kohatake S (1989) Estimation of the evapotranspiration rate by Thornthwaite or Harmon equation. Annu Disaster Prev Res Inst Kyoto Univ 32(2):309–317

Takahashi J, Onda Y, Hihara D, Tamura K (2018) Six-year monitoring of the vertical distribution of radiocesium in three forest soils after the Fukushima Dai-ichi nuclear power plant accident. J Environ Radioact 192:172–180

Tatsuno T, Hamamoto S, Nihei N, Nishimura T (2022) Vertical migration of cesium in weathered granite soil under flowing water condition depending on Cs concentration and states of dissolved organic matter. J Environ Manage 306:114409

Turner NB, Ryan JN, Saiers JE (2006) Effect of desorption on colloid-facilitated transport of contaminants: cesium, strontium, and illite colloids. Water Resour Res 42:W12S09. https://doi.org/10.1029/2006WR004972

Yamaguchi N, Takata Y, Hayashi K, Ishikawa S, Kuramata M, Eguchi S, Yoshikawa S, Sakaguchi A, Asada K, Wagai R, Makino T, Akahane I, Hiradat S (2012) Behavior of radiocaesium in soil-plant systems and its controlling factor: a review. Bull Natl Inst Agro Environ Sci 31:75–129

Chapter 15
Effect of Exchangeable and Nonexchangeable Potassium in Soil on Cesium Uptake by *Quercus serrata* Seedlings

Riona Kobayashi, Masaya Masumori, Takeshi Tange, Keitaro Tanoi, Natsuko I. Kobayashi, and Satoru Miura

15.1 Introduction

Following the accident at the Fukushima Daiichi Nuclear Power Plant (FNDPP) in 2011, radiocesium (^{134}Cs, ^{137}Cs) spread widely, contaminating forests in Fukushima Prefecture and surrounding prefectures. Unfortunately, ^{137}Cs has a long half-life of 30.2 years, and because forests circulate nutrients through biogeochemical cycles, it is difficult for radionuclides to flow out of these systems (Yamaguchi et al. 2012). Therefore, contamination by ^{137}Cs will be a challenge for the forest industry for a long time (IAEA 2002). The log production industry for cultivating Shiitake mushrooms suffered some of the greatest damage caused by the Fukushima disaster. The Forestry Agency in Japan has ruled that logs containing radiocesium above an index value of 50 Bq kg^{-1} should not be distributed to markets in order to ensure that the radiocesium concentration in Shiitake mushrooms grown on the logs is below the standard limit for general food in Japan; 100 Bq kg^{-1} (Ministry of Agriculture, Forestry and Fisheries 2012).

Konara oak (*Quercus serrata* Murray) was cultivated for mushroom log production in some areas of Fukushima Prefecture and its surrounding prefectures, but

R. Kobayashi
Graduate School of Agricultural and Life Sciences, The University of Tokyo, Tokyo, Japan

Present Address: Sumitomo Forestry Co. Ltd., Tokyo, Japan

M. Masumori (✉) · T. Tange · K. Tanoi · N. I. Kobayashi
Graduate School of Agricultural and Life Sciences, The University of Tokyo, Tokyo, Japan
e-mail: masumori@fr.a.u-tokyo.ac.jp

S. Miura
Center for Forest Restoration and Radioecology, Forestry and Forest Products Research Institute, Tsukuba, Ibaraki, Japan

© The Author(s) 2023
T. M. Nakanishi, K. Tanoi (eds.), *Agricultural Implications of Fukushima Nuclear Accident (IV)*, https://doi.org/10.1007/978-981-19-9361-9_15

shipments of logs are still stopped, because their radiocesium concentrations may exceed the index value. To resume log production in these areas, reducing the radiocesium concentration in Konara oak trees is an urgent issue. Suppressing radiocesium uptake by roots could prevent contamination of seedlings to be planted in the future and new shoots grown from stumps.

One possible solution to reduce root uptake of radiocesium is to increase the amount of potassium (K) in the soil. Plants take up cesium (Cs), because K transporters, especially under low K, can transport Cs, which has similar chemical properties (Nieves-Cordones et al. 2017; Qi et al. 2008). Therefore, higher concentrations of potassium ions (K^+) in the rhizosphere compete more effectively with Cs^+, so that the transporter preferentially absorbs K^+, thereby reducing absorption of Cs^+ (Fujimura et al. 2014; Zhu and Smolders 2000). The effect of high K^+ concentrations on Cs uptake suppression was also observed in hydroponically cultivated Konara oak seedlings (Kobayashi et al. 2019). Similarly, a pot experiment confirmed that Konara oak seedlings treated with K had lower Cs concentrations than untreated control pots (Kobayashi, unpublished). Consistent with these reports, outdoor K fertilization experiments using cypress (*Chamaecyparis obtusa*) seedlings and coconut palms (*Cocos nucifera* L.) confirmed that K addition to the soil suppresses uptake of radiocesium (Komatsu et al. 2017; Robison et al. 2009). As for K in soil, many studies have defined K extracted with 1 M ammonium acetate solution as exchangeable K and analyzed its suppression of Cs uptake. When exchangeable K concentration in soil is high, plants grown there absorb less radiocesium (Yamamura et al. 2018). Addition of K fertilizer such as potassium chloride to paddies can suppress the transfer of radiocesium to rice plants even when the original exchangeable K concentration of the soil is low (Fujimura et al. 2013; Kato et al. 2015).

On the other hand, there was reportedly no relation between the exchangeable K concentrations and ^{137}Cs uptake at several investigated sites (Kubo et al. 2018). Similarly, for Konara oak, a preliminary analysis of several-year-old native seedlings and rhizosphere soils throughout Japan showed no significant correlation between exchangeable K concentrations in soils and the transfer factor of stable Cs ([133]Cs), indicating the degree of Cs uptake. Therefore, in addition to exchangeable K, the influence of nonexchangeable K has attracted attention. Nonexchangeable K is a form that is held between layers of clay minerals such as mica, or is incorporated into the structure of the minerals. When there is no relationship between exchangeable K and Cs uptake, the effect of nonexchangeable K should be taken into account to better explain Cs uptake. Eguchi et al. (2015) and Ogasawara et al. (2019) used sodium tetraphenylborate (TPB) and boiling nitric acid (HNO_3) respectively, as methods to extract K from soils. While no correlation was found between exchangeable K concentration and ^{137}Cs uptake of rice (*Oryza sativa* L.), it was shown that the concentration of K extracted by these methods correlated with Cs uptake. These findings indicate that in Konara oak, not only exchangeable K, but also nonexchangeable K could be involved in suppressing Cs uptake.

In this study, we analyzed seedlings of Konara oak and their rhizospheric soils in order to clarify effects of nonexchangeable K as well as exchangeable K in determining Cs uptake by Konara oak. Hydrogen peroxide (H_2O_2), boiling HNO_3, and

TPB were used to extract K from soils, and the relationship between these K concentrations and Cs uptake by Konara oak was examined. Because the three methods extract different amounts of K from different fractions of soils, we tried to find the optimal K extraction method to evaluate Cs uptake of Konara oak. In this study, we collected samples not only in Fukushima Prefecture, but also in other uncontaminated areas; thus, we measured ^{133}Cs instead of radiocesium.

15.2 Materials and Methods

15.2.1 Sampling of Native Seedlings and Soils

We used pairs of current-year shoots of native Konara oak seedlings and soils around their roots collected from 36 sites in nine prefectures from February 7th to April 19th, 2017 (Fig. 15.1). Sampling sites were not limited to mountain forests, but were widely distributed in places where acorns of Konara oak were thought to have grown naturally: 1–5-year-old seedlings less than waist high were collected. Soils were collected at three points within a radius of 30 cm from the seedlings to a depth of 10 cm after completely removing litter. These three samples were mixed. The number of sample replicates ranged from 1 to 5, mostly 3, for each site; thus, the total number of pairs of shoots and soils we collected came to 106. Collected current-year shoots and soils were dried in a 70 °C dryer. Dried shoots were cut into pieces 5 mm in length and dried soils were passed through a 2-mm sieve. Carbon and nitrogen in soil samples were determined by the dry combustion method (SUMIGRAPH NC-22F, Sumika Chemical Analysis Service, Tokyo, Japan).

15.2.2 Extraction of K and ^{133}Cs from Soil

Potassium was extracted from soil samples using the four methods. Exchangeable K was extracted with 1 M ammonium acetate for 1 h at a soil:solution ratio of 1:10. To extract K derived from organic matter, 30 wt% H_2O_2 solution, adjusted to pH 2.0, was added to 1 g of soil, and heated at 85 °C for 8 h. After cooling, 10 mL of 1 M ammonium acetate was added to yield a supernatant after 1 h of shaking. This extract was designated as H_2O_2-K. Then, the residue was subsequently subjected to boiling HNO_3 decomposition. One molar HNO_3 was added to the residue and the mixture was heated at 115 °C for 20 min after boiling. After allowing the solution to cool, the supernatant was taken, and 10 mL of 100 mM HNO_3 was added to the residue to obtain another supernatant. This operation was performed again, and the three supernatants were mixed. This extract was identified as HNO_3-K. Potassium extraction using TPB was performed according to Carey et al. (2011), and the extract was labeled TPB-K. The concentration of exchangeable K was measured using an inductively coupled plasma optical emission spectrometer (ICP-OES) (Optima

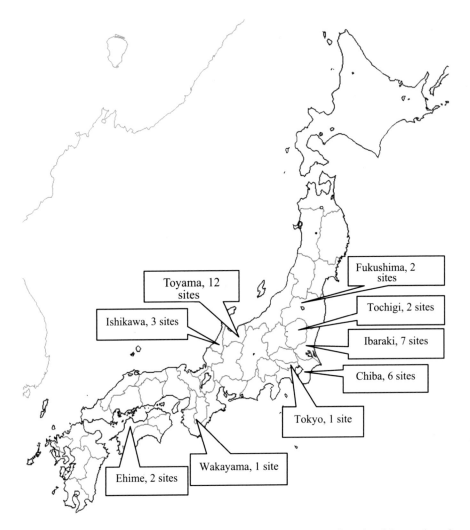

Fig. 15.1 Prefectures in which Konara oak seedlings and soils were collected and the number of sample sites in each prefecture

7300 DV, PerkinElmer, Waltham, Massachusetts), and concentrations of H_2O_2-K, HNO_3-K, and TPB-K were measured using atomic absorption spectrophotometry (Z-6100, Hitachi High-Technologies Corporation, Tokyo, Japan). In addition, ^{133}Cs concentration in the solution extracted with HNO_3 was measured using an inductively coupled plasma mass spectrometer (ICP-MS) (NexION 350, PerkinElmer, Waltham, Massachusetts), and was designated as HNO_3-Cs.

15.2.3   Decomposition of Current-Year Shoots and Measurement of ^{133}Cs Concentrations

To determine cesium and potassium concentrations in current-year shoots, about 0.3 g of shoots was digested with 10 mL of HNO_3 (60%) using a microwave digester (Multiwave 3000, Anton Parr, Graz, Austria) for 60 min. Concentrations of ^{133}Cs and K in the extracts were measured using ICP-MS and ICP-OES, respectively. The transfer factor (TF), which describes the extent of radiocesium transfer from soil to plants, is usually expressed as the content of radiocesium in the plant (Bq kg^{-1}) divided by the concentration of radiocesium in the soil (Bq kg^{-1}). In this study, however, we analyzed ^{133}Cs, and total ^{133}Cs concentration was not measured, because a dedicated facility was required for the measurement. Therefore, TF was defined by the following equation, and the ^{133}Cs uptake of Konara oak was evaluated using this equation. The reason for using the concentration of HNO_3-Cs is to calculate TF with a value close to the ^{133}Cs concentration of the whole soil, such as the TF of radiocesium, and because Cs cannot be extracted by the TPB extraction method.

$$TF = \frac{^{133}Cs \, \text{concentration in the seedling} \, (\mu g \, g^{-1})}{HNO_3\text{-Cs concentration in the soil} \, (\mu g \, g^{-1})}.$$

15.3   Results

15.3.1   Relationship Between Soil K Concentrations and TF

Mean values of TF were calculated for each of the 36 sampling sites, revealing a wide variation with a maximum value of 1.5 and a minimum value of 0.04. Figure 15.2 shows the relationship between each K concentration and TF. Each plot represents the mean and standard deviation at each sampling site. Correlation coefficients (r) are shown in the figures. The correlation coefficient between logarithmic TF and logarithmic exchangeable K concentration was −0.28, and it was nonsignificant (Fig. 15.2a). Relationships among logarithmic TF, logarithmic H_2O_2-K concentration, and logarithmic HNO_3-K concentration were significant ($p < 0.05$). The correlation coefficients between them were larger than that of the exchangeable K concentration: −0.41 for logarithmic H_2O_2-K, −0.51 for logarithmic HNO_3-K (Fig. 15.2b, c). On the other hand, the correlation coefficient between logarithmic TPB-K concentration and logarithmic TF was −0.29, and the correlation was not statistically significant (Fig. 15.2d).

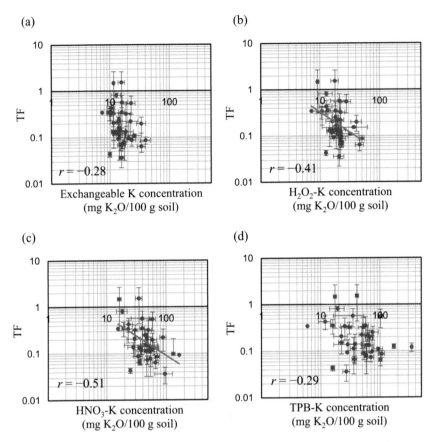

Fig. 15.2 Relationships between soil K concentration extracted by each method and ^{133}Cs transfer factor (TF) of Konara oak. (**a**) Exchangeable K. (**b**) H$_2$O$_2$-K. (**c**) HNO$_3$-K. (**d**) TPB-K. Each plot represents the mean ± SD for each sampling site. r is the correlation coefficient of the two variables. Grey lines indicate that the correlation coefficient is significant ($p < 0.05$)

15.3.2 Relationships Between Soil K Concentrations

The exchangeable K concentration ranged to 7.4 from 40.0 mg K$_2$O/100 g soil for 36 values, averaged by sampling site for 106 soils. Relationships between exchangeable K concentration and H$_2$O$_2$-K, HNO$_3$-K, and TPB-K concentrations are shown in Fig. 15.3. Exchangeable K concentration and H$_2$O$_2$-K concentration had the strongest correlation among all combinations, with a correlation coefficient of 0.91 (Fig. 15.3a). The H$_2$O$_2$-K concentration was distributed in the same range as the exchangeable K concentration, compared to the HNO$_3$-K and TPB-K concentrations. The slope of the regression coefficient was 1.2, indicating that the concentration of H$_2$O$_2$-K was 20% greater than the exchangeable potassium concentration, on average.

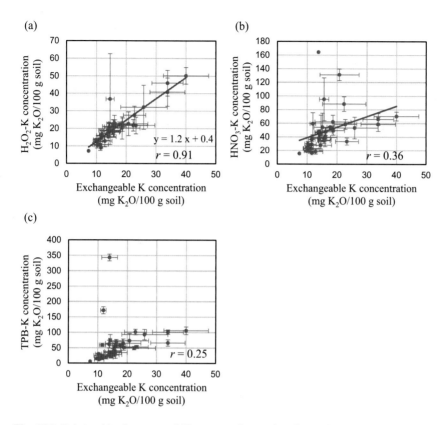

Fig. 15.3 Relationships between soil K concentrations and exchangeable K concentrations. (**a**) H_2O_2-K. (**b**) HNO_3-K. (**c**) TPB-K. Each plot represents the mean ± SD for each sampling site. r is the correlation coefficient of the two variables. Grey lines indicate that the correlation coefficient is significant ($p < 0.05$)

The correlation coefficient between exchangeable K and HNO_3-K was 0.36. In the range of HNO_3-K concentrations below 70 mg K_2O/100 g soil, a linear relationship can be observed with exchangeable K, but the site with higher HNO_3-K concentration was not on that line (Fig. 15.3b), which weakened the correlation of the two variables.

The TPB-K concentration tended to increase as the exchangeable K concentration increased at most sites, but a few sites showed very high TPB-K concentrations, despite the low exchangeable K concentrations (Fig. 15.3c). The correlation coefficient between these variables was 0.25. Figure 15.4 shows the relationship between HNO_3-K concentration and TPB-K concentration, both of which are often used to evaluate nonexchangeable K. From this figure, we found that the site having the highest HNO_3-K concentration and the site with the highest TPB-K concentration were not the same. Excluding these sites, the two variables were generally positively correlated.

Fig. 15.4 Relationship between soil HNO₃-K concentration and TPB-K concentration. Each plot represents the mean ± SD for each sample site. *r* is the correlation coefficient of the two variables

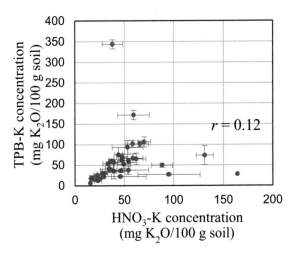

Fig. 15.5 Relationship between soil HNO₃-Cs concentration and ^{133}Cs content in current-year shoots of Konara oak. Each plot represents the mean ± SD for each sample site. *r* is the correlation coefficient of the two variables

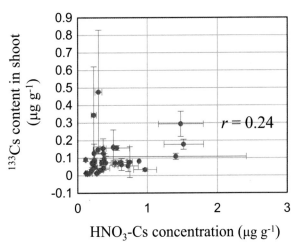

15.3.3 Soil HNO₃-Cs Concentration and Its Relationship with ^{133}Cs Contents in Shoots

Soil HNO₃-Cs concentrations ranged from a maximum of 1.5 μg g^{-1} to a minimum of 0.11 μg g^{-1}, a 15-fold variation. On the other hand, the ^{133}Cs content of current-year shoots of Konara oak showed a maximum value of 0.48 μg g^{-1} and a minimum value of 0.01 μg g^{-1}, which was about 50 times different. Figure 15.5 shows the relationship between HNO₃-Cs concentration in soils and ^{133}Cs content in shoots. There was no significant correlation between the two variables.

15.3.4   Relationships Between Soil K Concentrations and K Contents of Shoots

Figure 15.6 shows the relationship between the K content in current-year shoots of Konara oak and K concentrations in soil. K concentrations in the soil ranged from 7.4 to 40 mg K_2O/100 g soil for exchangeable K and from 6.0 to 343 mg K_2O/100 g soil for TPB-K, whereas the K content in current-year shoots did not differ much, ranging from 2.1 to 3.9 (Fig. 15.6). When the exchangeable K concentration and H_2O_2-K concentration were high, the K content in current-year shoots tended not to be low, but there were sites in which the K content of shoots was still high, even when the soil K concentration was low.

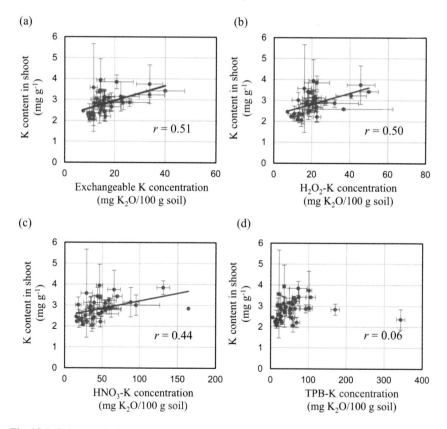

Fig. 15.6 Relationships between each soil K concentration and K content in current-year shoots of Konara oak. (**a**) H_2O_2-K. (**b**) HNO_3-K. (**c**) TPB-K. Each plot represents the mean ± SD for each sample site. r is the correlation coefficient of the two variables. Grey lines indicate that the correlation coefficient is significant ($p < 0.05$)

15.3.5   Relationship Between TF and Soil Carbon and Nitrogen Contents

Soil carbon content ranged from 0.80 to 16.9% and nitrogen content ranged from 0.04 to 1.11%. Figures 15.7 and 15.8 show the relationships between TF and carbon or nitrogen contents, respectively. Except for the two points that specifically showed large TF values, there was no site at which TF was high and carbon or nitrogen content was low, but no clear correlation was observed.

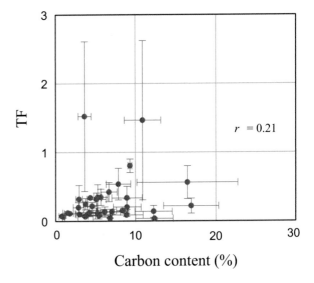

Fig. 15.7 Relationship between soil carbon content and ^{133}Cs transfer factor (TF) of Konara oak. Each plot represents the mean ± SD for each sample site. r is the correlation coefficient of the two variables

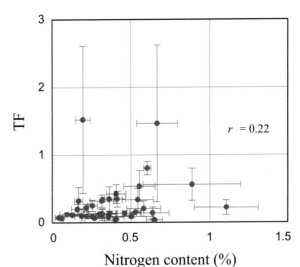

Fig. 15.8 Relationship between soil nitrogen content and ^{133}Cs transfer factor (TF) of Konara oak. Each plot represents the mean ± SD for each sample site. r is the correlation coefficient of the two variables

15.4 Discussion

Correlations between TF for ^{133}Cs of current-year shoots and K concentrations in soils extracted by four methods revealed that exchangeable K concentration, which is most commonly used for predicting TF, had little correlation with TF. On the other hand, the concentration of HNO$_3$-K, which extracts nonexchangeable K as well as exchangeable K, had the strongest correlation with TF. Therefore, in Konara oak, ^{133}Cs uptake is thought to be strongly affected not only by exchangeable K, which readily dissolves in soil solution to be available to plants, but also by nonexchangeable K that is sandwiched between clay mineral layers or incorporated into the mineral structure.

In this study, the correlation between TF and TPB-K concentration was weaker than that with HNO$_3$-K concentration. TPB can extract nonexchangeable K such as HNO$_3$, but TPB nondestructively extracts K between clay mineral layers to mimic the K uptake process by plants (Cox et al. 1999), while some K is extracted by destroying the mineral structure when using HNO$_3$ (Li et al. 2015). Although TPB is generally considered to extract more K than HNO$_3$ (Moritsuka 2009), Fig. 15.4 shows that the concentrations of TPB-K and HNO$_3$-K do not always match. Thus, the strength of the correlation with TF differed between HNO$_3$-K concentration and TPB-K concentration, because the extraction source and the extraction amount of two methods are different. Although some studies have measured soil nonexchangeable K concentrations using either HNO$_3$ or TPB and showed that they explained TF better than exchangeable K concentrations (Eguchi et al. 2015; Ogasawara et al. 2019), no previous studies have used multiple methods to extract nonexchangeable K and none have examined their relationships with TF of stable cesium. Although the relationship between HNO$_3$-K concentration and TF of current-year shoots of Konara oak was highly variable among samples, we expect that the multiple extraction methods employed in this study will improve the accuracy of predicting TF of Konara oak by characterizing adsorption of potassium in soil.

The variability of ^{133}Cs content in current-year shoots of Konara oak was larger than the variability of HNO$_3$-Cs concentration in soil, suggesting that ^{133}Cs uptake in Konara oak may be affected by factors other than just soil ^{133}Cs concentration (Fig. 15.5). This is because there was no correlation between HNO$_3$-Cs concentration in the soil and ^{133}Cs content in current-year shoots. Similarly for radiocesium, ^{137}Cs uptake is independent of soil ^{137}Cs concentration (Kanasashi et al. 2020).

Focusing on the relationship between K concentrations in soils and K contents of current-year branches of Konara oak, the variation in K content of the shoots was very small compared to the wide range of K concentrations in the soil, which varied by one to two orders of magnitude. This suggests that the function of exchangeable or nonexchangeable K in the soil on the uptake of K and Cs by Konara oak differs between K, which is an essential element, and Cs, which is not. No clear relationship between TF and either carbon or nitrogen content in the soil was found, suggesting that the effect of soil nutrient status on Cs uptake is small.

Chemical forms of soil K are diverse, influenced by clay mineral composition. Correspondingly, even within the range of the same species, *Quercus serrata*, differences in Cs uptake and transfer range from one to three orders of magnitude, depending on the environment. In order to improve the accuracy of predicting future radiocesium concentrations in Konara oak, further analysis should focus attention on nonexchangeable K, which has been little studied to date.

Acknowledgements We are grateful to Dr. Shoichiro Hamamoto and Mr. Yusuke Akiyama, Graduate School of Agricultural and Life Sciences, The University of Tokyo, for their guidance on soil analytical methods. We express our gratitude to all those who helped in collecting the samples, especially Mr. Tomoyuki Maruyama, Tochigi Prefecture Forest Research Center, Mr. Fukuda Kensuke, Ibaraki Prefectural Government Forestry Technology Center, Mr. Masami Iwasawa, Chiba Prefectural Agriculture and Forestry Research Center, Forestry Research Institute, Mr. Haruki Nakajima, Toyama Prefectural Agricultural, Forestry and Fisheries Research Center, Forest Research Institute, Mr. Jiro Kodani, Ishikawa Agricultural and Forestry Research Center, Forestry Experiment Station, Ms. Yumiko Yamashita, Wakayama Prefectural Forestry Experiment Station, Mr. Nobuyuki Toyota, Ehime University, Faculty of Agriculture, Dr. Kazunori Shimada, Forestry and Forest Products Research Institute, Tama Forest Science Garden, and Dr. Tatsuya Tsurita and Dr. Tsutomu Kanasashi, Forestry and Forest Products Research Institute.

References

Carey PL, Curtin D, Scott CL (2011) An improved procedure for routine determination of reserve-K in pastoral soils. Plant Soil 341:461–472. https://doi.org/10.1007/s11104-010-0658-x

Cox AE, Joern BC, Brouder SM, Gao D (1999) Plant-available potassium assessment with a modified sodium tetraphenylboron method. Soil Sci Soc Am J 63:902–911. https://doi.org/10.2136/sssaj1999.634902x

Eguchi T, Ohta T, Ishikawa T, Matsunami H, Takahashi Y, Kubo K, Yamaguchi N, Kihou N, Shinano T (2015) Influence of the nonexchangeable potassium of mica on radiocesium uptake by paddy rice. J Environ Radioact 147:33–42. https://doi.org/10.1016/j.jenvrad.2015.05.002

Fujimura S, Yoshioka K, Saito T, Sato M, Sato M, Sakuma Y, Muramatsu Y (2013) Effects of applying potassium, zeolite and vermiculite on the radiocesium uptake by rice plants grown in paddy field soils collected from Fukushima Prefecture. Plant Prod Sci 16:166–170. https://doi.org/10.1626/pps.16.166

Fujimura S, Ishikawa J, Sakuma Y, Saito T, Sato M, Yoshioka K (2014) Theoretical model of the effect of potassium on the uptake of radiocesium by rice. J Environ Radioact 138:122–131. https://doi.org/10.1016/j.jenvrad.2014.08.017

IAEA (2002) Modelling the migration and accumulation of radionuclides in forest ecosystems—report of the Forest Working Group of the Biosphere Modelling and Assessment (BIOMASS) Programme Theme 3. Iaea-Biomass. IAEA, Vienna, pp 1–127

Kanasashi T, Miura S, Hirai K, Nagakura J, Itô H (2020) Relationship between the activity concentration of ^{137}Cs in the growing shoots of *Quercus serrata* and soil ^{137}Cs, exchangeable cations, and pH in Fukushima, Japan. J Environ Radioact 220:106276. https://doi.org/10.1016/j.jenvrad.2020.106276

Kato N, Kihou N, Fujimura S, Ikeba M, Miyazaki N, Saito Y, Eguchi T, Itoh S (2015) Potassium fertilizer and other materials as countermeasures to reduce radiocesium levels in rice: results of urgent experiments in 2011 responding to the Fukushima daiichi nuclear power plant accident. Soil Sci Plant Nutr 61:179–190. https://doi.org/10.1080/00380768.2014.995584

Kobayashi R, Kobayashi NI, Tanoi K, Masumori M, Tange T (2019) Potassium supply reduces cesium uptake in Konara oak not by an alteration of uptake mechanism, but by the uptake competition between the ions. J Environ Radioact 208:106032. https://doi.org/10.1016/j.jenvrad.2019.106032

Komatsu M, Hirai K, Nagakura J, Noguchi K (2017) Potassium fertilisation reduces radiocesium uptake by Japanese cypress seedlings grown in a stand contaminated by the Fukushima Daiichi nuclear accident. Sci Rep 7:1–10. https://doi.org/10.1038/s41598-017-15401-w

Kubo K, Hirayama T, Fujimura S, Eguchi T, Nihei N, Hamamoto S, Takeuchi M, Saito T, Ota T, Shinano T (2018) Potassium behavior and clay mineral composition in the soil with low effectiveness of potassium application. Soil Sci Plant Nutr 64:265–271. https://doi.org/10.1080/00380768.2017.1419830

Li T, Wang H, Zhou Z, Chen X, Zhou J (2015) A nano-scale study of the mechanisms of non-exchangeable potassium release from micas. Appl Clay Sci 118:131–137. https://doi.org/10.1016/j.clay.2015.09.013

Moritsuka N (2009) Forms of potassium in agricultural soils in Japan: evaluation at regional, field and root-zone scales. Jpn J Soil Sci Plant Nutr 80:80–88

Nieves-Cordones M, Mohamed S, Tanoi K, Kobayashi NI, Takagi K, Vernet A, Guiderdoni E, Périn C, Sentenac H, Véry A-A (2017) Production of low-Cs$^+$ rice plants by inactivation of the K$^+$ transporter OsHAK1 with the CRISPR-Cas system. Plant J 92:43–56. https://doi.org/10.1111/tpj.13632

Ogasawara S, Eguchi T, Nakao A, Fujimura S, Takahashi Y, Matsunami H, Tsukada H, Yanai J, Shinano T (2019) Phytoavailability of ^{137}Cs and stable Cs in soils from different parent materials in Fukushima, Japan. J Environ Radioact 198:117–125. https://doi.org/10.1016/j.jenvrad.2018.12.028

Qi Z, Hampton CR, Shin R, Barkla BJ, White PJ, Schachtman DP (2008) The high affinity K$^+$ transporter AtHAK5 plays a physiological role in planta at very low K$^+$ concentrations and provides a caesium uptake pathway in Arabidopsis. J Exp Bot 59:595–607. https://doi.org/10.1093/jxb/erm330

Robison WL, Brown PH, Stone EL, Hamilton TF, Conrado CL, Kehl S (2009) Distribution and ratios of ^{137}Cs and K in control and K-treated coconut trees at Bikini Island where nuclear test fallout occurred: effects and implications. J Environ Radioact 100:76–83. https://doi.org/10.1016/j.jenvrad.2008.10.016

Yamaguchi N, Takata Y, Hayashi K, Ishikawa S, Kuramata M, Eguch S, Yoshikawa S, Sakaguchi A, Asada K, Wagai R, Makino T, Akahane I, Hiradate S (2012) Behavior of radiocaesium in soil-plant systems and its controlling factor: a review. Bull Natl Inst Agro Environ 31:75–129

Yamamura K, Fujimura S, Ota T, Ishikawa T, Saito T, Arai Y, Shinano T (2018) A statistical model for estimating the radiocesium transfer factor from soil to brown rice using the soil exchangeable potassium content. J Environ Radioact 195:114–125. https://doi.org/10.1016/j.jenvrad.2018.04.026

Zhu YG, Smolders E (2000) Plant uptake of radiocaesium: a review of mechanisms, regulation and application. J Exp Bot 51:1635–1645. https://doi.org/10.1093/jexbot/51.351.1635

Chapter 16
Ten-Year Transition of Radiocesium Contamination in Wild Mushrooms in the University of Tokyo Forests After the Fukushima Accident

Toshihiro Yamada

16.1 Introduction

Radiocesium released from the Fukushima Daiichi nuclear power plant (FDNPP) accident on March 11, 2011, spread over a wide area of East Japan. Wild mushrooms often contain a high level of radiocesium even in areas with lower levels of contamination. The University of Tokyo has seven research forests located in East Japan, 250–660 km from FDNPP, where radiocesium contamination is relatively low. Some varieties of mushrooms collected there, however, contained radiocesium over the regulatory level of 100 Bq/kg in 2011 over the following years. Surveys of radiocesium contamination of wild mushrooms were conducted in the University of Tokyo Forests (UTFs), because fungi, including mushrooms, are a major components of the forest ecosystem.

Mushrooms are known to accumulate radiocesium (Byrne 1988; Kammerer et al. 1994; Mascanzoni 1987; Muramatsu et al. 1991; Sugiyama et al. 1990, 1994). However, the radiocesium concentration ratio in mushrooms relative to the soil was rather low (Heinrich 1992). This trend was also observed in a previous study (Yamada 2019). Further, a considerable proportion of ^{137}Cs in forest soil is retained by the fungal mycelia, and fungi are thought to prevent the elimination of radiocesium from ecosystems (Brückmann and Wolters 1994; Guillitte et al. 1994; Vinichuk and Johanson 2003; Vinichuk et al. 2005). Thus, fungal activity is likely to contribute substantially to the long-term retention of radiocesium in the organic layers of forest soil by recycling and retaining radiocesium between fungal mycelia

T. Yamada (✉)

The University of Tokyo Chichibu Forest, Graduate School of Agricultural and Life Sciences, The University of Tokyo, Chichibu, Saitama, Japan
e-mail: yamari@uf.a.u-tokyo.ac.jp

T. M. Nakanishi, K. Tanoi (eds.), *Agricultural Implications of Fukushima Nuclear Accident (IV)*, https://doi.org/10.1007/978-981-19-9361-9_16

185

and soil (Muramatsu and Yoshida 1997; Steiner et al. 2002; Yoshida and Muramatsu 1994, 1996).

The dynamics of radiocesium in the ecosystem can be inferred by comparing the natural decay according to the physical half-life of radiocesium with actual changes, that is, the changes owing to the biological or ecological half-life considering migration, absorption, and excretion. According to analyses from that point of view, the ecological half-life of radiocesium in mushrooms has been reported to be longer than that of plants (Kiefer et al. 1996; Zibold et al. 2001; Strandberg 2004; Fielitz et al. 2009).

Previous studies (Yamada 2013, 2019) have summarized radiocesium contamination of wild mushrooms in UTFs after the Fukushima accident. We found rapid uptake of radiocesium in one species of mushroom after the Fukushima accident. In addition, we found residual contamination from the global fallout of atmospheric nuclear weapons tests or the Chernobyl accident. We also attempted to analyze the factors that determine the changes in radiocesium concentration (Yamada 2018, 2019). In the current study, the dynamics of radiocesium were surveyed over a 10 year period, and features of the time course transition in mushroom contamination were summarized.

16.2   Research Sites and Sampling

Mushrooms that appeared from the ground every Autumn from 2011 to 2020 were collected. In addition, their presumptive soil substrates, that is, the O horizon (organic litter layer, called the A_0 horizon in Japan), the A horizon (mineral layer and accumulated organic matter), and the C/O horizon [mineral layer with a small quantity of organic matter, which is relatively unaffected by pedogenic processes (Soil Survey Staff 2014)], were also collected. Collected mushroom species were shown in Yamada et al. (2018) and Yamada (2019). Samples from the following three research forests were used for the current analyses.

Chichibu: The University of Tokyo Chichibu Forest (UTCF)

Fuji: Fuji Iyashinomori Woodland Study Center (FIWSC)

Chiba: The University of Tokyo Chiba Forest (UTCBF)

The location of each research forest, examples of mushrooms, the appearance of the environment where samples were collected, and sample preparation for radioactivity measurement can be seen in Yamada (2019). The radioactivity concentrations of ^{134}Cs and ^{137}Cs were determined using a germanium semiconductor detector. The distribution of radiocesium deposition and γ-ray air dose rate in East Japan in 2011 was presented in the previous studies (Yamada 2013, 2019).

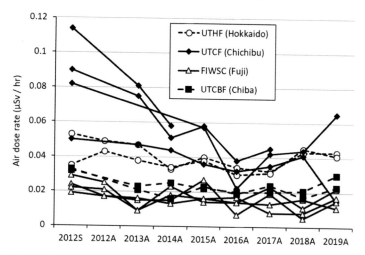

Fig. 16.1 Change in air dose rate 1 m above ground at each University of Tokyo Forest site. *S* spring, *A* autumn

16.3   Gamma Ray Air Dose Rate at the Mushroom Collection Sites (Fig. 16.1)

Gamma ray air dose rate (μSv/h) 1 m above ground level was measured with a dose rate meter using a CsI (Tl) scintillation detector (Yamada 2019). Although considerable variation in dose rate was observed among UTFs due to environmental variation such as geological features and contamination level, trends of changes and levels in dose rate were similar within each UTF. The air dose rate in Hokkaido (The University of Tokyo Hokkaido Forest, UTHF), where no contamination derived from the Fukushima accident was recognized, is also shown in Fig. 16.1 as a control. Similar levels of pre-Fukushima contamination from global fallout were estimated from ^{137}Cs/^{134}Cs ratio in soils in Chiba (UTCBF) and Fuji (FIWSC). The air dose rate before the Fukushima accident in Fuji, however, was lower than in Chiba, probably owing to geological features. The dose rate slightly decreased in Chiba with time, whereas the decrease was not clear in Fuji. Since 2015, the dose rate in Fuji has become almost similar to the dose rate recorded in Chiba. Although contamination due to the Fukushima accident did not reach Hokkaido, the dose rate was higher in Hokkaido (UTHF) than that in Fuji and Chiba. One year after the Fukushima accident, the dose rate in Chichibu (UTCF) was higher than that in other UTFs, and was over 0.1 μSv/h particularly in high mountainous areas, and then gradually reduced by approximately half by 2015. Although it seems that some hotspots remain in Chichibu, the dose rate has dropped to approximately 0.05 μSv/h since 2016. The dose rates at all UTFs sites are considered to be stable at almost preaccident values.

16.4   Dynamics of Radiocesium Contamination in the University of Tokyo Forests

16.4.1   Overall Trends (Fig. 16.2)

There were some patterns in the changes in ^{137}Cs concentration (Yamada 2019; Yamada et al. 2019). In some cases, such as in Fuji's saprotrophic fungus *Pholiota lubrica*, the ^{137}Cs concentration tended to decrease clearly and consistently. In contrast, it has been reported in Europe that the concentration of radiocesium in

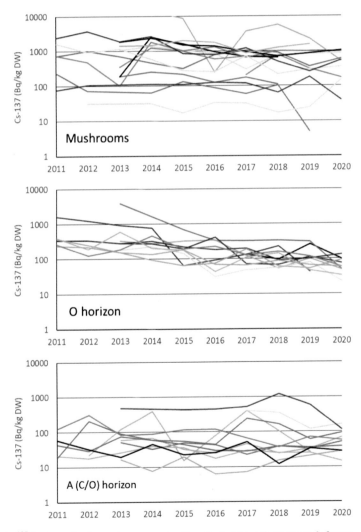

Fig. 16.2 ^{137}Cs dynamics in mushrooms and soils collected from three research forests

mushrooms increased for a few years after the Chernobyl accident (Borio et al. 1991; Smith and Beresford 2005). A similar pattern was seen in several cases, such as mycorrhizal fungus *Suillus grevillea*, showing a tendency to increase from 2011 to 2012 and then decrease. However, the ^{137}Cs concentration often did not clearly change. In this way, the changes in ^{137}Cs concentration in mushrooms were roughly seen to follow three distinct patterns. In the O horizon, there were many sites where the ^{137}Cs concentration gradually decreased, but there were also sites where the decreasing tendency was not clear. In the C/O horizon observed in Fuji or A horizon, the ^{137}Cs concentration increased once, but in many cases no significant change was observed. The changes up to 2020 are the same as the previously reported trends up to 2015 and 2017, and the ^{137}Cs concentration tends to gradually decreases in the O horizon and tends to be retained in the A or C/O horizon. Mushrooms showed a variety of tendencies in the middle of litter and soil.

16.4.2 Trend at the Same Sampling Sites (Fig. 16.3)

Yearly changes in ^{137}Cs concentration were compared between the mushrooms, O horizon, A (or C/O) horizon. In the O horizon, there was a clear tendency for the ^{137}Cs concentration to decrease over time. It decreased in any period even if it was divided into 1–4, 4–7, and 7–10 years after the accident. A large difference in the temporal change of ^{137}Cs concentration among mushrooms was observed, and a variety of patterns were seen, such as those that decreased, those that did not

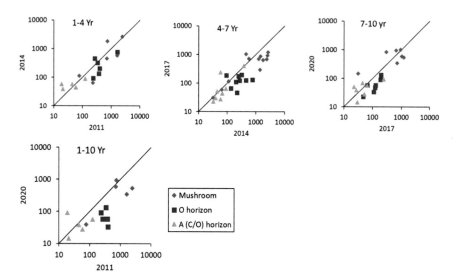

Fig. 16.3 Scatter diagram indicating ^{137}Cs (Bq/kg DW) changes from 2011 to 2020 in mushrooms and soils from the same sampling sites. The oblique solid line ($Y = X$) indicates the same ^{137}Cs concentration between 2011 and 2020

decrease, and those that increased temporarily. In general, the mushrooms tended to retain ^{137}Cs stably from the beginning, or alternatively to decrease partially followed by retention. In the A or C/O horizon, an increase or decrease of radiocesium was observed, but there was no clear continuous increase or decrease. It seems that part of the ^{137}Cs in the O horizon has moved to the A or C/O horizon. These trends in mushrooms and soil continued for 10 years after the accident.

16.5 Examples of Radiocesium Transfer at the Same Sampling Sites (Fig. 16.4)

The transition of radiocesium derived from the Fukushima accident can be elucidated when decay correction is performed for [134]Cs. However, there are drawbacks to this method, such as falling below the detection limit and large error due to the rapid decrease in its concentration. Therefore, we determined the changes in ^{137}Cs decay corrected for the date of the Fukushima accident, March 11, 2011. Here, as typical examples of changes observed, trends in ^{137}Cs concentrations at one site in Chichibu and two sites in Fuji are shown.

At the site that seems to be a local hotspot in Chichibu, the concentration of ^{137}Cs in the O horizon was relatively high, but the subsequent decrease was also remarkable. It is thought that temporary imports of radiocesium into the A horizon occurred; however, the export became dominant after that. The ^{137}Cs in the mycorrhizal *Russula emetica* in this location fluctuates according to the concentration in the O horizon and the A horizon.

Contamination from Fukushima was slight in Fuji. However, the saprotrophic fungus *P. lubrica*, which is said to have a superficial hyphal layer, had already absorbed a large amount of radiocesium by the fall of 2011 and showed a tendency of a gradual decrease in the radiocesium level. Radiocesium did not change significantly in the mycorrhizal fungus *Lactarius laeticolor* and it continued to retain these levels. Mycorrhizal *Lactarius hatsudake* and *Suillus luteus* tended to predominantly discharge radiocesium after initial absorption. The absolute amount of contamination in the O horizon was not large in Fuji, but there was a tendency for export of radiocesium from the O horizon, and a tendency for retention or import to be greater in the C/O horizon.

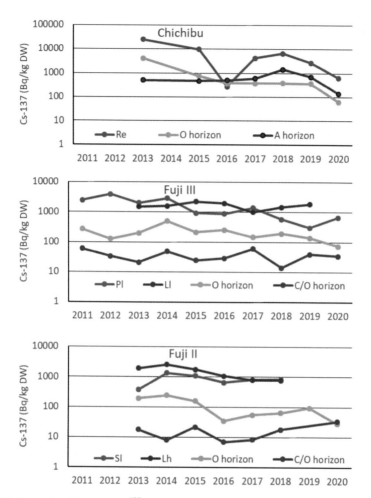

Fig. 16.4 Examples of dynamics of ^{137}Cs, corrected for March 11, 2011, at the same sampling site. *Re Russula emetica, Pl Pholiota lubrica, Ll Lactarius laeticolor, Sl Suillus luteus, Lh Lactarius hatsudake*

16.6   Transfer of Radiocesium—Changes in Decay Corrected ^{137}Cs/^{134}Cs Radioactivity Ratio in Each University Forest (Fig. 16.5)

Changes in ^{134}Cs concentration decay corrected for Mar. 11, 2011, indicate contamination only from Fukushima. To match the transfer of radiocesium from global fallout and Fukushima, a simple comparison of the decay corrected ^{134}Cs and ^{137}Cs values is difficult to interpret. Therefore, decay correction was performed and changes in ^{137}Cs/^{134}Cs were compared (Rühm et al. 1997).

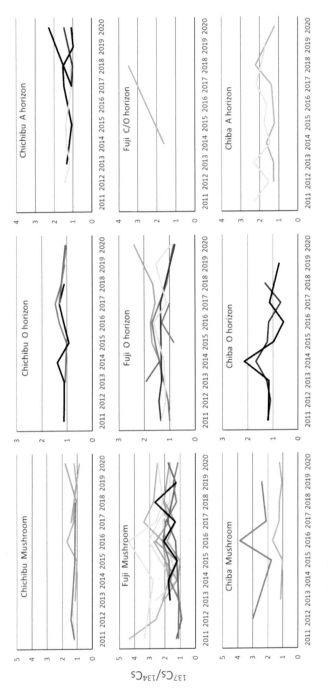

Fig. 16.5 Changes in $^{137}Cs/^{134}Cs$ ratio, corrected for March 11, 2011, in each of three UTFs

High values of ^{137}Cs/^{134}Cs were observed in mycorrhizal mushrooms in 2011, which indicates that radiocesium from before the Fukushima accident remained and accumulated in the fungal mycelium (Yamada 2013, 2019). After that, this value was maintained in some cases, but also decreased in many cases. Many of the imports and exports were balanced, but imports exceeded exports in some cases. Therefore, this decrease of ^{137}Cs/^{134}Cs was considered to be caused by the absorption of radiocesium derived from the Fukushima accident. For saprotrophic mushrooms, it was initially low at almost 1.0, but then gradually increased. This suggested that although the amount of radiocesium remaining in the mycelia before the Fukushima accident was small, radiocesium derived from Fukushima was taken up by the time of collection (7 months after the accident) and then released and transferred to the outside. The value in the O horizon ranged from 1 to 2, indicating that ^{137}Cs before Fukushima remained in the surface layer. Since ^{134}Cs was scarcely present in the C/O horizon, changes in ^{137}Cs/^{134}Cs could not be tracked. However, changes in ^{137}Cs in the previous section suggested the export or import of radiocesium depending on the location.

In Chichibu's mycorrhizal mushrooms, this ^{137}Cs/^{134}Cs value was low, that is, the radiocesium derived from Fukushima was absorbed quickly, and the contribution of the previous radiocesium was small. This value was not high even in the O and A horizon, suggesting the relatively high contamination from the Fukushima accident. The cases studied from Chiba were also of mycorrhizal mushrooms. One of them had a large value, and the value was maintained every year throughout the period. In the A horizon of Chiba, the value was initially higher than 1. This showed that ^{137}Cs before Fukushima remained approximately the same as ^{137}Cs derived from Fukushima, and it is presumed that it was incorporated into mushrooms.

16.7   Transfer of Radiocesium (^{137}Cs) Derived from Fukushima Accident Compared with Pre-Fukushima ^{137}Cs (Fig. 16.6)

To understand the biological and ecological dynamics such as migration and retention of radiocesium, the contribution of the Fukushima accident and global fallout due to past nuclear weapons tests or the Chernobyl accident was evaluated from the ratio of ^{134}Cs and ^{137}Cs. Furthermore, since the proportion of ^{137}Cs derived from Fukushima is considered to be greatly affected by the degree of contamination, the transition of the relationship between the concentration of ^{137}Cs and the proportion of ^{137}Cs before Fukushima and derived from Fukushima was investigated for each sample collected at the same sites (Fig. 16.6).

In Fig. 16.6, the change to the lower right indicates that ^{137}Cs derived from Fukushima is taken in, the change to the upper left indicates that ^{137}Cs derived from Fukushima is preferentially discharged. The change to the left horizontal direction

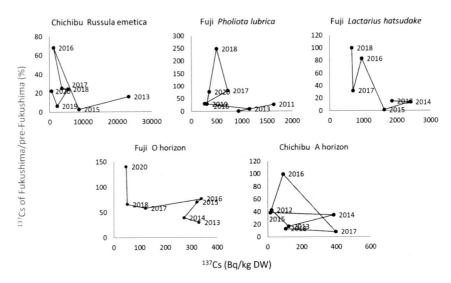

Fig. 16.6 Examples of changes in the ratio of ^{137}Cs derived from the Fukushima accident compared with pre-Fukushima ^{137}Cs

shows that ^{137}Cs from before Fukushima and ^{137}Cs from Fukushima are discharged in the same way.

Various patterns of changes were seen in the mushrooms, but many of them, such as *L. hatsudake* and *S. luteus*, changed from the upper left to the lower right and then changed to the upper left again. Another pattern that changed from the right to the left was seen in *R. emetica* in Chichibu and *P. lubrica* in Fuji. Intermediate forms between the two were also seen relatively frequently. In the case of *L. hatsudake* and *S. luteus*, it is speculated that ^{137}Cs derived from Fukushima was gradually absorbed after the accident, and then as the ^{137}Cs from Fukushima in soil such as the O horizon decreased, the emission of ^{137}Cs from Fukushima exceeded the absorption. In the case of *P. lubrica*, a large amount of ^{137}Cs derived from Fukushima was already absorbed in 2011, but the subsequent changes in ^{137}Cs seen in this species are similar to those seen in mycorrhizal mushrooms.

In the O horizon, the pattern that changed from the lower right to the upper left is conspicuous, indicating that the Fukushima-derived ^{137}Cs preferentially migrated to the outside. Changes in the A or C/O horizon often did not show any particular pattern.

These changes were not only because radiocesium is well retained in mushrooms, but the radiocesium released from the Fukushima accident was exported from the O horizon relatively quickly, whereas the radiocesium from before Fukushima generally remained in the O horizon. In other words, the radiocesium remaining in the O horizon is not adsorbed by clay minerals and should be able to easily migrate, but most of it tends to be retained in the material cycle system of the soil surface layer. Although it cannot be evaluated quantitatively, we presume that the retention of radiocesium in the hyphae is one factor for this. However, as in the case of *R. emetica* and *P. lubrica*, the proportion of ^{137}Cs derived from the Fukushima accident has tended to decrease in mushrooms in recent years. The radiocesium value in the ecosystem from Fukushima may have now attained values similar to those seen before the Fukushima accident.

16.8   Concluding Remarks

We have measured the radioactivity concentrations of ^{134}Cs and ^{137}Cs over a monitoring period of 10 years, and analyzed and clarified the characteristics of radioactivity in wild mushrooms found in the University of Tokyo Forests. Monitoring data of ^{134}Cs and ^{137}Cs concentration provided the time course patterns of radiocesium accumulation, transfer, and selective retention of absorbed radiocesium in mushrooms. Further long-term monitoring of radiocesium is necessary to more precisely determine the dynamics of the contamination, through analyses comparing the levels of ^{134}Cs and ^{137}Cs; this is generally considered difficult because of the short half-life of ^{134}Cs.

Acknowledgements   I sincerely thank the staff of the University of Tokyo Forests for collecting and preparing the samples, and Drs. N. I. Kobayashi, K. Tanoi and T. M. Nakanishi for measuring the radioactivity in samples and for their valuable comments.

References

Borio R, Chiocchini S, Cicioni R, Esposti PD, Rongoni A, Sabatini P, Scampoli P, Antonini A, Salvadori P (1991) Uptake of radiocesium by mushrooms. Sci Total Environ 106:183–190

Brückmann A, Wolters V (1994) Microbial immobilization and recycling of ^{137}Cs in the organic layers of forest ecosystems: relationship to environmental conditions, humification and invertebrate. Sci Total Environ 157:249–256

Byrne AR (1988) Radioactivity in fungi in Slovenia, Yugoslavia, following the Chernobyl accident. J Environ Radioact 6:177–183

Fielitz U, Klemt E, Strebl F, Tataruch F, Zibold G (2009) Seasonality of ^{137}Cs in roe deer from Austria and Germany. J Environ Radioact 100:241–249

Guillitte O, Melin J, Wallberg L (1994) Biological pathways of radionuclides originating from the Chernobyl fallout in a boreal forest ecosystem. Sci Total Environ 157:207–215

Heinrich G (1992) Uptake and transfer factors of ^{137}Cs by mushrooms. Radiat Environ Biophys 31: 39–49

Kammerer L, Hiersche L, Wirth E (1994) Uptake of radiocaesium by different species of mushrooms. J Environ Radioact 23:135–150

Kiefer P, Pröhl G, Müller G, Lindner G, Drissner J, Zibold G (1996) Factors affecting the transfer of radiocaesium from soil to roe deer in forest ecosystems of southern Germany. Sci Total Environ 192:49–61

Mascanzoni D (1987) Chernobyl's challenge to the environment: a report from Sweden. Sci Total Environ 67:133–148

Muramatsu Y, Yoshida S (1997) Mushroom and radiocesium. Radioisotopes 46:450–463

Muramatsu Y, Yoshida S, Sumiya M (1991) Concentrations of radiocesium and potassium in basidiomycetes collected in Japan. Sci Total Environ 105:29–39

Rühm W, Kammerer L, Hiersche L, Wirth E (1997) The ^{137}Cs/^{134}Cs ratio in fungi as an indicator of the major mycelium location in forest soil. J Environ Radioact 35:129–148

Smith JT, Beresford NA (2005) Radioactive fallout and environmental transfers. In: Smith JT, Beresford NA (eds) Chernobyl–catastrophe and consequences. Springer, Berlin

Soil Survey Staff (2014) Keys to soil taxonomy, 12th edn. Natural Resources Conservation Service, Washington, DC

Steiner M, Linkov I, Yoshida S (2002) The role of fungi in the transfer and cycling of radionuclides in forest ecosystems. J Environ Radioact 58:217–241

Strandberg M (2004) Long-term trends in the uptake of radiocesium in *Rozites caperatus*. Sci Total Environ 327:315–321

Sugiyama H, Iwashima K, Shibata H (1990) Concentration and behavior of radiocesium in higher basidiomycetes in some Kanto and the Koshin districts, Japan. Radioisotopes 39:499–502

Sugiyama H, Shibata H, Isomura K, Iwashima K (1994) Concentration of radiocesium in mushrooms and substrates in the sub-alpine forest of Mt. Fuji Japan. J Food Hyg Soc Jpn 35:13–22

Vinichuk MM, Johanson KJ (2003) Accumulation of ^{137}Cs by fungal mycelium in forest ecosystems of Ukraine. J Environ Radioact 64:27–43

Vinichuk MM, Johanson KJ, Rosén K, Nilsson I (2005) Role of the fungal mycelium in the retention of radiocaesium in forest soils. J Environ Radioact 78:77–92

Yamada T (2013) Mushrooms: radioactive contamination of widespread mushrooms in Japan. In: Nakanishi TM, Tanoi K (eds) Agricultural implications of the Fukushima nuclear accident. Springer, Tokyo

Yamada T (2018) Radiocesium dynamics in wild mushrooms in the University of Tokyo Forests after the Fukushima nuclear accident. Water Sci 364:81–99

Yamada T (2019) Radiocesium dynamics in wild mushrooms during the first 5 years after the Fukushima accident. In: Nakanishi TM, O'Brien M, Tanoi K (eds) Agricultural implications of the Fukushima nuclear accident (III). Springer, Tokyo

Yamada T, Omura K, Saito T, Igarashi Y, Takatoku K, Saiki M, Murakawa I, Iguchi K, Inoue M, Saito H, Tsuji K, Kobayashi NI, Tanoi K, Nakanishi TM (2018) Radiocesium contamination of wild mushrooms collected from the University of Tokyo Forests over a 6-year period (2011–2016) after the Fukushima nuclear accident. Misc Inf Univ Tokyo For 60:31–47

Yamada T, Murakawa I, Iguchi K, Oomura K, Igarashi Y, Saito T, Takatoku K, Saito H, Kobayashi NI, Tanoi K, Nakanishi TM (2019) Radiocesium dynamics in wild mushrooms in the University of Tokyo Forests during the first 7 years after the Fukushima nuclear accident. Kanto J For Res 70:81–84

Yoshida S, Muramatsu Y (1994) Accumulation of radiocesium in basidiomycetes collected from Japanese forests. Sci Total Environ 157:197–205

Yoshida S, Muramatsu Y (1996) Environmental radiation pollution of fungi. Jpn J Mycol 37:25–30

Zibold G, Drissner J, Kaminski S, Klemt E, Miller R (2001) Time-dependence of the radiocaesium contamination of roe deer: measurement and modeling. J Environ Radioact 55:5–27

Chapter 17
Challenge to Resume Production of Mushroom Bed Logs by Potassium Fertilizer Application

Masaya Masumori, Natsuko I. Kobayashi, Keitaro Tanoi, Naoto Nihei, Satoru Miura, and Tsutomu Kanasashi

17.1 Potential Effect of Forest Management to Reduce Radiocesium Content in Hardwood Trees

Radiocesium that adhered to the surfaces of tree trunks and branches immediately following the nuclear power plant accident has gradually diminished as bark peels off the trees. However, some radiocesium still remained on trunk and branch surfaces even several years after the accident (Fig. 17.1). Cutting these trees to induce regeneration of new shoots from stumps (coppicing) might be one way to reduce radiocesium contamination, since the radiocesium concentration in new

This chapter is a translation, with modifications, of Section III-2 of a Japanese publication, "Resumption of Use and Restoration of Shiitake Mushroom Log Forests in Radioactively Contaminated Areas" by the Forestry and Forest Products Research Institute, 2018, ISBN:978-4-905304-92-0.

M. Masumori (✉) · N. I. Kobayashi · K. Tanoi
Graduate School of Agricultural and Life Sciences, The University of Tokyo, Tokyo, Japan
e-mail: masumori@fr.a.u-tokyo.ac.jp

N. Nihei
Department of Agriculture, Fukushima University, Fukushima, Japan

S. Miura
Center for Forest Restoration and Radioecology, Forestry and Forest Products Research Institute, Tsukuba, Ibaraki, Japan

T. Kanasashi
Center for Forest Restoration and Radioecology, Forestry and Forest Products Research Institute, Tsukuba, Ibaraki, Japan

Present Address: Institute of Environmental Radioactivity, Fukushima University, Fukushima, Japan

T. M. Nakanishi, K. Tanoi (eds.), *Agricultural Implications of Fukushima Nuclear Accident (IV)*, https://doi.org/10.1007/978-981-19-9361-9_17

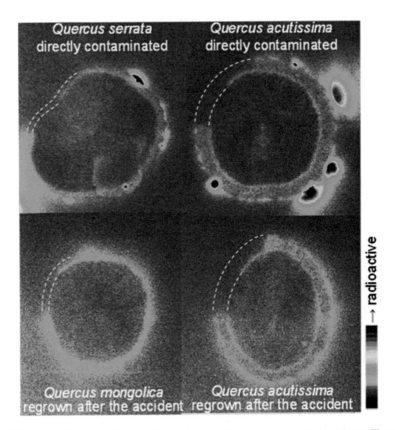

Fig. 17.1 Autoradiographs of cross sections of oak trunks grown in a contaminated area. The upper left portion of each section, indicated by dashed lines, was partially stripped of its bark. Strongly radioactive spots can be seen in sections of the trunk where radioactive materials were directly deposited at the time of the accident (top two cross sections). Radioactivity was also found in the bark of trunks that sprouted from stumps after the accident (bottom two cross sections)

trunks is assumed to be lower than that of directly contaminated trunks (unpublished). Another option would be to plant new seedlings, instead of coppicing. Although new seedlings would take years longer to achieve harvestable size compared to regenerating forests by coppicing, producing trees with lower concentrations of radiocesium should be prioritized. The source of contamination for both coppiced shoots and newly planted seedlings is radiocesium remaining in the soil that can be absorbed by roots. Thus, in order to produce bed logs for shiitake mushrooms in forests where radiocesium has been deposited, it is important to minimize absorption of radiocesium as much as possible.

17.2   Negative Correlation Between Exchangeable Potassium in Soil and the Radiocesium Concentration in New Shoots

It has been reported that radiocesium content in current-year shoots correlates well with trunk radiocesium content (Kanasashi et al. 2020). Accordingly, we analyzed how much radiocesium was transferred to current-year shoots in hardwood stands in various regions of Fukushima. We found that there was a large variation in the amount of radiocesium transferred to new shoots in hardwood forests, even if the degree of soil contamination was about the same. Forests with larger contents of exchangeable potassium in the soil commonly have lower transfer factors of radiocesium to the shoots (Fig. 17.2). In many crops, radiocesium absorption by roots is suppressed when potassium is abundant around the roots (Yamaguchi et al. 2016). The results of this study indicate that potassium has a similar effect in woody plants such as Konara oak.

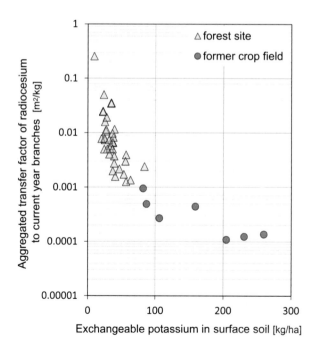

Fig. 17.2 The relationship between exchangeable potassium in soil and aggregated transfer factors to current year branches. Higher potassium content in the soil resulted in lower amounts of radiocesium transferred from the soil through roots to branches of the current year. Aggregated transfer factors were calculated by adding data from former crop fields to current-year branch data (Kanasashi et al. 2020)

In forest stands with abundant exchangeable potassium in the soil, such as a former crop field that had been continuously fertilized, it should be possible to grow trees that have absorbed little radiocesium.

17.3   Application of Potassium Fertilizer to Increase Exchangeable Potassium in the Soil

We applied potassium chloride fertilizer to the soil surface to see if we could increase the amount of exchangeable potassium. After 1 year, the amount of exchangeable potassium near the surface, where root density is generally high, increased in most forests, although some forests showed no increase (Fig. 17.3). We presumed this was due to differences in soil type. A comparison of radiocesium concentrations absorbed by newly planted seedlings over the course of a year showed a trend toward lower concentrations in forests where potassium fertilizer had been applied (Fig. 17.4). Potassium fertilizer likely reduces absorption of radiocesium by newly planted seedlings.

However, 2 years after application of potassium fertilizer of several tens of grams per square meter (= several hundred kg/ha), the content of exchangeable potassium

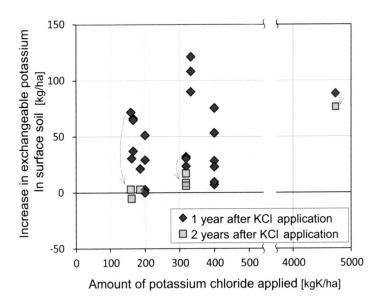

Fig. 17.3 Increase in potassium content in surface soil 1 year after application of potassium fertilizer to the surface. At most study sites, potassium chloride fertilizer application to the soil surface increased the exchangeable potassium content in the surface soil after 1 year, compared to before the application, but the potassium content returned to the original level after 2 years if the amount of potassium fertilizer applied was small

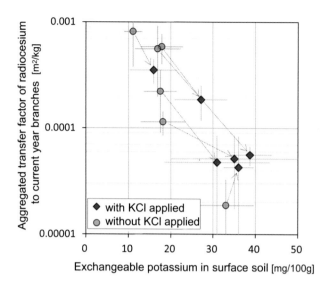

Fig. 17.4 Reduction of radioactivity in newly planted seedlings by application of potassium fertilizer to the soil surface. When potassium chloride fertilizer was applied to the soil surface (◇), exchangeable potassium in the soil increased and less radiocesium was transferred to branches of newly planted seedlings than in an adjacent plot, where no fertilizer was applied (○). However, no reduction effect was observed at sites that originally had high potassium levels (the lowest point)

began to return to the original level (Fig. 17.3). To sustain the potassium fertilizer effect until harvest, it may be necessary to apply larger amounts or to repeat the application of potassium fertilizer.

Investigations at these experimental fertilization sites are ongoing and time-course results will be reported in the future.

Acknowledgements This study was funded by research grants from the Bio-oriented Technology Research Advancement Institution, NARO, Japan (a research program for development of innovative technology; 28028C), and JSPS KAKENHI Grant Number JP 17K11950.

References

Kanasashi T, Miura S, Hirai K, Nagakura J, Itô H (2020) Relationship between the activity concentration of ^{137}Cs in the growing shoots of *Quercus serrata* and soil ^{137}Cs, exchangeable cations, and pH in Fukushima, Japan. J Environ Radioact 220:106276. https://doi.org/10.1016/j.jenvrad.2020.106276

Yamaguchi N, Taniyama I, Kimura T, Yoshioka K, Saito M (2016) Contamination of agricultural products and soils with radiocesium derived from the accident at TEPCO Fukushima Daiichi Nuclear Power Station: monitoring, case studies and countermeasures. Soil Sci Plant Nutr 62: 303–314. https://doi.org/10.1080/00380768.2016.1196119

Chapter 18
Studies on the Revitalization of Radioactive-Contaminated Mushroom Log Forests: Focus on Shoots

Satoru Miura

18.1 Introduction

The radioactive contamination of forests caused by the accident at the TEPCO's Fukushima Daiichi Nuclear Power Plant has had various effects on people's lives and livelihood in the satoyama areas of Fukushima Prefecture. The forest products are diverse both in variety and for use. Even if the amount of radioactive materials that fell because of the nuclear power plant accident and the intensity of their radiation are the same, the type of forest products used and how they were used has brought a significantly different impact.

The radioactive contamination of the forests resulted in high air dose rates, which hindered activities such as logging, afforestation, and other forestry work and recreation. This is a serious matter; there are still areas designated as difficult-to-return zones (evacuation zones). The long-term effects on the recovery of daily life and livelihood are enormous. In contrast, the use of forest trees and other forest products contaminated by radioactive materials has been restricted according to the radioactivity levels of the forest products, which also have a significant impact on people's lives and livelihood. The severity of the impact determined by the radioactivity concentration of the forest products is more complicated than in the case of air dose rates. The overall picture of such radioactive contamination of forests, forestry industry, and the livelihood of people living in satoyama is described in detail in a book based on published data (Hashimoto et al. 2022). To summarize the effects of radioactive contamination on the forestry industry sector, low regulation values of radiocesium activity concentration were set for edible forest products and raw materials for foodstuff production, relatively high index values were set for other

S. Miura (✉)
Center for Forest Restoration and Radioecology, Forestry and Forest Products Research Institute, Tsukuba, Ibaraki, Japan
e-mail: miura@ffpri.affrc.go.jp

cases, or no regulation values (index values) were set for building materials. Under these circumstances, more than 10 years after the nuclear accident, one of the most important issues for the forestry and forest industry in Fukushima is the resumption of full-scale hardwood forest operations (Miura 2016; Hashimoto et al. 2022). In particular, there is a strong desire among forestry business entities in the contaminated areas to seek a pathway to resume the use of hardwood forests for mushroom log production.

18.2 What Is Needed to Resume Hardwood Mushroom Log Forestry in the Abukuma Region?

The Fukushima Daiichi Nuclear Power Plant is located on the eastern side of Fukushima Prefecture facing the Pacific Ocean. Its western hinterland, called Abukuma Highlands, is a gently undulating mountainous area composed of granite, stretching 50 km from east to west and 150 km from north to south. Before the nuclear power plant accident, this area formed one of Japan's largest mushroom log production areas and supplied many logs outside of Fukushima Prefecture (Miura 2016). However, because of radioactive contamination caused by the nuclear power plant accident, the bed logs and logs for mushroom production were contaminated soon after the accident, and the production and shipments were completely halted across the area. Even now, no one in the Abukuma region has resumed full-scale production of logs as in the past. Only some pulp chips for paper manufacturing are being shipped, and experimental log production is being conducted.

When hardwoods are used as mushroom logs, the greatest concern is the concentration of future trunks used for bed logs. Immediately after the nuclear power plant accident, the Forestry Agency conducted an emergency survey to determine the actual situation regarding the contamination of the logs and bed logs for mushroom production and the radioactive cesium concentration in the produced mushrooms. Based on this survey, in March 2012, the Forestry Agency tentatively determined and noticed index values of 50 Bq/kg for mushroom logs and 200 Bq/kg for sawdust, which can be used for mushroom bed cultivation, and provided guidance to log and sawdust producers and mushroom farmers (Hashimoto et al. 2022). Each producer has hesitated to decide whether to continue log production, because they do not have a clear idea of how many years they should wait before they can resume log production in the hardwood forests where production is currently halted. In other words, forest owners and forestry business entities are faced with the necessity to decide whether to continue the log production business or withdraw from log production. The predicted radiocesium concentration at the time of future harvesting is needed for this purpose.

The Miyakoji Office of Fukushima Central Forestry Association, located in Miyakoji Town, Tamura City, in the Abukuma region, was known as a producer of high-quality mushroom logs (Fig. 18.1a, b). In the satoyama of this area,

Fig. 18.1 (a) A mushroom log forest and paddy fields in early summer around the satoyama area in Abukuma Highland at Miyakoji Town, Tamura City, Fukushima. (b) A mushroom log forest in winter at Miyakoji Town

hardwood trees once used as firewood and charcoal forests were no longer being used owing to the energy revolution. In the Miyakoji area, the Forestry Association took the initiative in converting the forests for firewood and charcoal to log forests for mushroom production in the 1970s. A hundred hectares or more of secondary hardwood forests were cut down each year in the 1990s and 2000s to produce and sell mushroom logs. The forests were cyclically utilized to mushroom log forests, whereby the target area was moved sequentially, returning to the same forest 20 years later to harvest again the coppices that had grown to a size where they could be used as logs (Fig. 18.2). The Forestry Association acted as a coordinator, and the entire community worked together to form sustainable coppicing forests. Under such circumstances, the nuclear power plant accident occurred in March 2011, and all production activities came to a halt. Harvesting forests take 20 years from the time they are cut down to the next harvest. To resume such cyclical sustainable forestry, which was cut off by the nuclear accident, reliable predictions must be made as to whether or not the trunk of mushroom log forests harvested and

Fig. 18.1 (continued)

regenerated now will have dropped below the index value of 50 Bq/kg in 20 years. Although the dynamics of radiocesium in forest ecosystems is complex, (1) the structure of the prediction equation for future forecasting (mechanism of radiocesium dynamics) must be understood, and (2) it must be clarified which of the explanatory variables in the prediction equation have a large contribution to the magnitude of fluctuation (i.e., clarifying the range of variation of the explanatory variables).

This book contains seven such reports on *Quercus serrata* (Konara oak) or mushroom log forests. After the Fukushima nuclear accident, the resumption of operations in mushroom log forests became a major forestry issue, but it remains unresolved. Studies on mushroom log forests initiated at the University of Tokyo (e.g., Kobayashi et al. 2019a, 2019b) have been vigorously pursued at the Forestry and Forest Products Research Institute (FFPRI; Kanasashi et al. 2020; Kenzo et al. 2020; Sakashita et al. 2021). This chapter introduces the latest major study results of those research groups. They attempted to promote the resumption of mushroom log forests and, in particular, explained the significance of focusing on current-year shoots in the study of coppice forests of Konara oak.

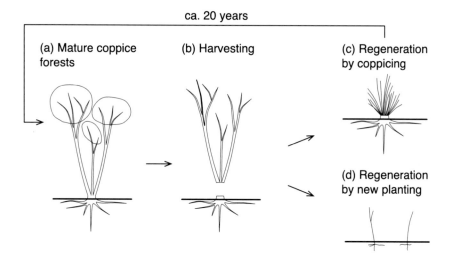

Fig. 18.2 Cycle of management for coppicing hardwood forest producing mushroom logs, repeated every 20 years

18.3 Overview of Radioactive Contamination in Forests: What We Know So Far

Surveys began immediately after the Fukushima nuclear power plant accident to determine the actual status of radioactive contamination of forests (Kato et al. 2012). According to the results of this prompt survey, the amount deposited on trees above ground exceeded 50% of the total amount of fallout. In addition, the Forestry Agency and national and prefectural agencies, such as the FFPRI and the Fukushima Prefecture, set up new test sites and began fixed-point observation after the nuclear accident (Hashimoto et al. 2022). The survey results revealed that radiocesium was translocated throughout the trees promptly after the onset of radioactive contamination. In August 2011, 5 months after the occurrence of radioactive contamination, the FFPRI surveyed major forest tree species, such as Japanese cedar, in three areas in Fukushima Prefecture, where the initial deposition differed by a hundredfold, to clarify the radioactive cesium concentration and deposition in the major components of the forest ecosystem (Kuroda et al. 2013; Komatsu et al. 2016; Imamura et al. 2017). The higher the radiocesium activity concentration in the wood inside the tree, the more contaminated the study site was; it was proportional to the amount of radiocesium deposited in the forest (Kuroda et al. 2013). Radiocesium had been taken up inside the trees and transported and diffused to a certain extent into the interior of the trunks as early as 5 months after the accident. There are two uptake routes of radiocesium by trees: absorption from the surface of the tree, such as leaves and bark, and absorption from the soil via the roots. Attempts have been made to

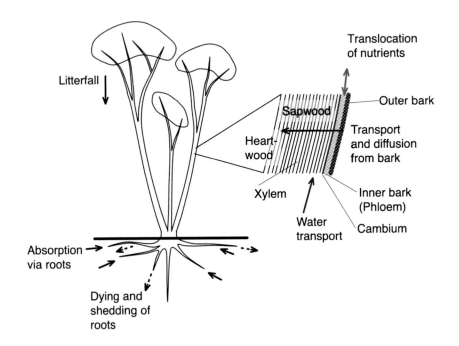

Fig. 18.3 Schematic overview of the uptake and transport of deposited radiocesium into and within trees

clarify the contribution rate of these two pathways (Imamura et al. 2021), but the contribution rate is considered to change over time and remains unclear.

The author has outlined the uptake of cesium by trees and its movement within the tree in Fig. 18.3. Tree trunks are covered with bark. Inside the bark is a thin cambium layer where cell division occurs. The bark is divided into two parts: the inner bark, which is living tissue and forms the phloem, and the outer bark, which is composed of dead cork-like tissue. Immediately after the nuclear power plant accident, it is believed that radioactive cesium adhering to the outer bark surface was absorbed into the tree through the bark surface. The phloem, which forms the inner bark, functions to translocate nutrients produced by the leaves to other parts of the tree, and sap flows through it. In contrast, a tissue called the xylem is formed on the inner side of the cambium. The xylem supports the tree and also functions as a water-conducting structure, transporting water absorbed from the roots to the leaves. Ionized radiocesium in the soil and litter is absorbed within the water.

Recent experimental studies using Japanese cedar have shown that radiocesium injected outside the Japanese cedar sapwood moves toward the center in the radial direction of the sapwood with a speed on the order of days by a combination of active transport and diffusion (Kuroda et al. 2020). Although this experiment was conducted on Japanese cedar, it is likely that radiocesium translocating through the phloem also moves faster than expected in the radial direction inside the trunk in other tree species, such as Konara oak.

The author and coresearchers conducted a whole-tree digging surveyes in 2014 in a coppiced Konara oal forest. The radiocesium in the above- and below-ground biomass of the Konara oak forest distributed as a result of reflecting such transport of cesium within the tree. Three years after the nuclear accident, radiocesium was detected in all parts of Konara oak, from fine roots to the central core of the root stump. An analysis of root data collected in radiocesium surveys of various forests after the Fukushima nuclear accident reported that the radiocesium activity concentrations in roots were relatively higher than those of the soil going down deeper in the soil (Sakashita et al. 2020). Sakashita et al. (2020) gathered that root mortality and shedding may be a driving force for the vertical downward movement of radiocesium in the soil.

These studies on the distribution and movement of radiocesium within trees suggested that radiocesium that fell in forests was incorporated into the movement of material cycles in forest ecosystems more quickly than initially expected and circulated within the forests. Such radiocesium distribution in forests and future radiocesium activity concentrations in trees have also been predicted by models since the earlier phase after the Fukushima nuclear accident (Hashimoto et al. 2013; Thiry et al. 2018). The models have been updated with the addition and enrichment of new observational data (Hashimoto et al. 2020; Thiry et al. 2020). Comparative studies among multiple models have been conducted (Hashimoto et al. 2021; Hashimoto 2023). Given the different responses of multiple models to the same data set, Hashimoto (2023) noted the importance of a better understanding of the root uptake ratio and the biogeochemical processes. The key to the appropriate use of models is to identify factors that significantly impact the radiocesium dynamics within the forest ecosystem and within trees.

18.4   Key Factors of Radiocesium Uptake in Mushroom Log Forests

As mentioned in the previous section, radiocesium in forests moved significantly between soil and trees weekly to monthly. Considering that 10 years have already passed since the fallout of radiocesium into the forest, it is expected that the radiocesium distribution in the trees has neared a state of equilibrium in the soil environment. The average harvesting cycle of hardwood forests, such as Konara oak, used for mushroom cultivation is approximately 20 years, more than half of which has already passed since the nuclear power plant accident occurred. The coppicing ability of hardwood trees decreases with age, so forest owners have to determine whether or not to continue the coppice forest operations by harvesting and coppicing regeneration of the trees, and the time limit for this decision is approaching.

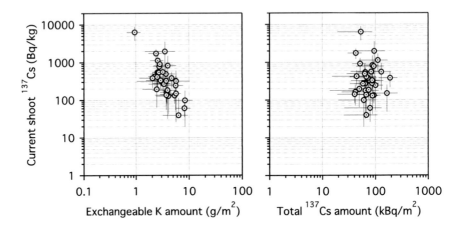

Fig. 18.4 Relationship between exchangeable potassium amount (left) or total ^{137}Cs amount (right) in the surface soil of 0–5 cm in depth and ^{137}Cs activity concentration in current-year shoots of Konara oak (*Quercus serrate*). (Source: Reproduced with data from Kanasashi et al. 2020)

The primary factor determining the degree of radioactive contamination of forest trees is first the amount of initial deposition of radioactive cesium. The second most important factor is how much radiocesium is absorbed by the trees from the soil. Kanasashi et al. (2020) clearly showed this by the 2016–2017 survey results of current-year shoots and soils at 34 sites of coppiced Konara oak forests that were harvested and regenerated after the nuclear accident. The mushroom log production by the Forestry Association in the Miyakoji area had completely stopped because of radioactive contamination caused by the nuclear accident. The study results (Kanasashi et al. 2020) in mushroom log forests in various locations in such an area showed a clear negative correlation between the amount of exchangeable potassium in forest soils at a depth of 0–5 cm and the radiocesium activity concentrations in current-year shoots of Konara oak on both logarithmic graph (Fig. 18.4 [left]). In contrast, the radiocesium activity concentration of current-year shoots did not correlate with the amount of ^{137}Cs in the soil (Fig. 18.4 [right]). Under the homogeneous initial deposition of radiocesium in the study area, the amount of exchangeable potassium was apparently the main factor determining radiocesium uptake by Konara oak rather than the radiocesium inventory. However, this does not mean that the initial deposition of radiocesium, which varied by a factor of 1000 in Fukushima Prefecture, does not affect radiocesium absorption. Undoubtedly, the degree of contamination determined by the amount of initial deposition is the primary factor in the amount of radiocesium absorbed by trees in the wider area. As mentioned earlier, in a survey conducted in summer 2011 in cedar forests with different degrees of contamination, the radiocesium concentration in the wood of trees was proportional to the degree of contamination (Kuroda et al. 2013). However, under the condition of being in an area with the same degree of radiocesium contamination as Miyakoji Town, the amount of exchangeable potassium in the soil has a stronger influence on radiocesium absorption than the slight difference in radiocesium amount in the soil.

18.5 Uptake Competition Between Cesium and Potassium

Plant roots absorb alkaline elements, such as potassium and cesium, through transporters and channels. Although cesium is not an essential element for plants, it is a homologous element of potassium. It causes strong absorption competition due to its close ionic radius (White and Broadley 2000; Zhu et al. 2000). Potassium is a major essential element for plants, and the plants actively try to absorb the necessary amount for growth. However, when there is not enough of it in the growth medium, the absorption of cesium, which has a similar chemical form to potassium, increases. Conversely, when there are sufficient potassium ions, cesium absorption is suppressed. This competition with potassium in cesium absorption by plants was known even before the Chernobyl accident (Zhu et al. 2000). Absorption suppression of cesium by potassium was also actively used as a measure against radiocesium contamination in crop production after the Fukushima nuclear accident. For example, testing exchangeable potassium concentrations in soil and applying potassium fertilizers were required for paddy rice cultivation in Fukushima Prefecture (Kato et al. 2015). This competition between cesium and potassium uptake has likely been caused by the amount of exchangeable potassium in forest soils where Konara oak grows (Kanasashi et al. 2020).

   In this book, the plant physiological characteristics of cesium absorption in trees are presented in Chap. 5, presenting the results of hydroponic trials conducted with Konara oak and rice as experimental material (Kobayashi et al. 2023, 2019a; Kobayashi and Kobayashi 2023). The competition between potassium and cesium ions also works in root uptake in Konara oak, as confirmed by hydroponic tests (Kobayashi et al. 2019b). However, the selective absorption of potassium observed in rice was not observed in Konara oak (Kobayashi et al. 2019a). Only competition for ion absorption was observed in Konara oak. In addition, in this hydroponic experiment, the difference in the uptake rate of cesium absorbed by leaves and branches was only twofold compared to a 60-fold difference in the potassium ion concentration in the culture medium. In contrast, the multipoint field surveys showed a 100-fold difference in radiocesium activity concentrations in Konara oak shoots against a tenfold difference in exchangeable potassium in the soil at 0–5 cm depth (Kanasashi et al. 2020). In the first place, it is not possible to directly compare the results of hydroponic experiments that examined the characteristics of plant physiological uptake of potassium and cesium ions by roots to the results of studies on the amount of exchangeable potassium in forest soils in the field. However, even taking this into account, it is highly probable that the differences in the amount of exchangeable potassium in forest soils significantly affect radiocesium absorption by Konara oak. This point should be further investigated in other areas to verify the universality of its nature. Furthermore, the possibility that factors other than ion competition, which has been clarified in hydroponic tests, may be involved must also be confirmed from various perspectives.

In contrast, the results of Kanasashi et al. (2020) showed the potential to be used now as a tool for finding available logging areas with low radiocesium absorption. Figure 18.4 (left) shows the relationship between soil exchangeable potassium contents and radiocesium activity concentrations of current-year shoots. However, if we focus only on the radiocesium activity concentration of current-year shoots, one important fact emerges. For a given soil level of exchangeable potassium, the average radiocesium activity concentration of current-year shoots is about one order of magnitude. For example, for an exchangeable potassium level of 20 g/m^2 per 0–5 cm depth, the radiocesium activity concentration of current-year shoots is distributed in the range of 200–2000 Bq/kg. At an exchangeable potassium level of 70 g/m^2, the radiocesium activity concentrations of current-year shoots range from 40 to 400 Bq/kg. Once the exchangeable potassium level is determined, the range of radiocesium activity concentration in current-year shoots is also determined correspondingly and does not deviate from it by more than two or three orders of magnitude. To predict and estimate radiocesium uptake by Konara oak, it is often assumed that it is necessary to know the amount of exchangeable potassium in the soil, which is the main factor. However, from a different point of view, if one examines the radiocesium activity concentrations of Konara oak shoots in a given forest area, they should fall within a range of about tenfold, no matter where one examines the radiocesium activity concentrations. Although the variation was large (~10 times), a bidirectional 1:1 relationship was observed between the amount of exchangeable potassium in the soil and the radiocesium concentration in current-year shoots of Konara oak. The logging companies need the radiocesium activity concentration in the logs used for mushroom production, not the amount of exchangeable potassium in the soil. Therefore, the results of Kanasashi et al. (2020) indicated that, considering only the practical purpose of determining whether or not a log forest can be used in the future, it is sufficient to examine only the radiocesium activity concentrations in current-year shoots. It is hoped that the development of a practical method to determine the forest area that will be established can be carried out in accordance with this concept, because it is easier to examine the radiocesium activity concentration in current-year shoots than the exchangeable potassium of the soil.

Studies on hardwood log forests focusing on current-year shoots of Konara oak are currently at this stage of development. Methods to predict radiocesium absorption and future radiocesium activity concentrations are still being developed. Further efforts are being made to improve the accuracy of predicting future radiocesium activity concentrations in hardwood forests for mushroom logs by combining the elucidation of mechanisms with the development of practical tools.

18.6 Distinguishing Current-Year Shoots

As mentioned above, the author's group has focused on current-year shoots, the growing parts of trees that can be easily investigated. Based on this assumption, Sakashita et al. (2021) conducted a detailed study of the seasonal variation of radiocesium activity concentrations in current-year shoots to extend the survey period in the field beyond the winter dormancy period. As a result, the radiocesium concentration in current-year shoots of Konara oak in the Miyakoji area increased rapidly with the opening of leaves in spring, then gradually decreased, and remained relatively stable from August or later (Sakashita et al. 2021). The radiocesium activity concentration of current-year shoots is the lowest and most stable during the dormant period from November after defoliation to April before the next year's leaf opening. Further analysis of the data obtained in these studies led to the proposal in Chap. 13 of this book of a method to estimate radiocesium activity concentrations of current-year shoots in dormant season from radiocesium activity concentrations of leaves (Sakashita et al. 2023). In addition, enabling surveys with leaves from summer to autumn would have an advantage in forest management planning, because the radiocesium absorption characteristics of the targeted forests can be evaluated before autumn, when harvesting and regeneration of log forests are at full-scale.

The following paragraphs will explain how to identify current-year shoots in the field, which are useful as an indicator of radiocesium absorption by trees.

Current-year shoots can be identified by careful observation of the shoots. Deciduous trees form axillary and apical buds at the base of leaves and the tips of shoots in the summer to autumn of the previous year, which are the source of growth for the following year's leaves and branches. After leaf fall, the axillary and apical buds remain on the shoot covered by bud scales, stop growing, and attach to the shoot over winter. These are called winter buds. In spring, winter buds swell and differentiate into leaves and branches (shoots), which begin to grow; in summer to autumn, winter buds form again on the shoots that have grown that year. Figure 18.5a illustrates a current-year shoot with leaves collected at the beginning of October and a current-year shoot after leaf fall collected in April. Whether or not a branch is a current-year shoot can be determined by how the leaves grow. A shoot with leaves attached directly to the branch is a current-year shoot (Fig. 18.5a, b). After the leaves fall in the autumn, winter buds in the middle or at the apex of the shoot can be used to determine current-year shoots. Some short current-year shoots may not have leaves or winter buds in the middle of the branch. However, if they are current-year shoots, winter buds (apical buds) will still form at the tips of the elongated shoots. Most winter buds on the current-year shoot will open and grow into leaves or shoots during the next growing season. Even if they do not open, they are no longer living buds, so if a dormant winter bud is dead, it is considered an old shoot from the previous year. Especially, because winter buds are more conspicuous on a well-grown branch, it is not so difficult to determine the current-year shoot, if actually experienced.

Fig. 18.5 (**a**) Identification of current-year shoots and previous year branches by leaves and winter buds (collected at the beginning of October). Source: The original figure was created by Masaya Masumori (The University of Tokyo). (**b**) Identification of primary, secondary, and tertiary growths within the current-year by winter buds and nodes (collected in the middle of April)

Like leaves, current-year shoots are the parts of the tree that grow during the current-year, and they accumulate nutrients and minerals absorbed from the soil. They are also relatively easy to identify along with leaves. However, there are some points to note when actually sampling in the field. During the growing season from spring to summer, current-year shoots may undergo a secondary or tertiary new

Fig. 18.5 (continued)

growth. In the Miyakoji area, where the survey was conducted, winter buds open and new leaves begin to develop around early May. After that, depending on the weather conditions, new shoots may grow again in July, August, or September, when the lingering summer heat is severe. This is called secondary growth. In the upper canopy of trees, even if the tip of a shoot is branched, there are often cases in which the root side of the branch is also a current-year shoot (Fig. 18.5b). Therefore, the determination of the current-year shoot should be based on the attachment of leaves or winter buds to the shoot, not just branching characteristics at the nodes. Because Kanasashi et al. (2020) did not distinguish among primary, secondary, and tertiary growth of current-year shoots, the variation in the relationship between soil exchangeable potassium and radiocesium activity concentrations in current-year shoots may be due in part to the mix of secondary and nonsecondary growth in current-year shoots. The effect of conducting current-year shoots survey without distinguishing between primary-grown and secondary- or tertiary-grown shoots on the uncertainty of radiocesium activity concentration of current-year shoots remains to be evaluated.

18.7 Measures by Potassium Fertilization to Reduce Radiocesium Absorption

Up to this point, studies have focused on the distribution of exchangeable potassium in forest soils and explained how to find usable forests among contaminated mushroom log forests. The idea is to use forests as they are to the extent possible without incurring costs, which can be considered a passive measure. This is an extension of the long-established relationship with forests cultivated by people living in the satoyama area, and it is a rational approach. In contrast, there is also an attempt to apply potassium fertilization to trees in forests, which was carried out on a large scale in Fukushima Prefecture to suppress the absorption of radiocesium in farmland.

In 2014, a potassium fertilization experiment was conducted when planting Japanese cypress (*Chamaecyparis obtusa*) seedlings in a radioactively contaminated forest, and the suppression effect of potassium on the absorption of radioactive cesium was confirmed (Komatsu et al. 2017). The author's group also conducted potassium fertilization experiments in coppice forests of Konara oak and reported the results in Chap. 17 of this book (Masumori et al. 2023). The year after fertilization, soil exchangeable potassium increased and radiocesium activity concentrations in current-year shoots of Konara oak decreased. However, a year later, the exchangeable potassium concentration decreased again and returned to the original level, and the radiocesium activity concentration of current-year shoots also increased again (Masumori et al. 2023). In contrast, in Sweden, a one-time application of potassium chloride fertilizer at 100 kg/ha by weight of potassium was tested in 1992 to suppress radiocesium absorption by trees, mosses, and fungi in shrubland contaminated by the Chornobyl nuclear accident (Rosén et al. 2011). A follow-up study conducted

17 years later until 2009 showed that radiocesium activity concentrations in trees, mosses, and fungi were significantly lower in the potassium-applied area than in the control area. In Japanese coniferous forests, the cycle extends at least 40–50 years until harvest, when trees are used for lumber forests, and even secondary hardwood forests for mushroom logs require 20 years. When potassium fertilization is applied to suppress radiocesium absorption in forests, it is necessary to carefully examine whether the application is sufficiently effective until the harvesting period and whether the radiocesium concentrations in logs and other products at harvest can be lowered to the target level.

As a study focusing on the adsorption characteristics of potassium and cesium on soil, four different extractants were used to extract potassium from soils in Chap. 15 of this book, and tests were also conducted on soil from areas outside of Fukushima Prefecture (Kobayashi et al. 2023). Different extractants were analyzed for nonexchangeable and exchangeable potassium to determine the relationship with the transfer coefficient of ^{133}Cs in current-year shoots of Konara oak, with the result that the concentration of nonexchangeable potassium extracted with thermal nitric acid had the highest correlation with the transfer coefficient (Kobayashi et al. 2023). In agricultural soils, the potassium extraction characteristics of soils have revealed that differences in clay mineral composition are an important factor affecting the effectiveness of potassium application (Eguchi et al. 2015). Cesium and potassium adsorption characteristics in forest soils are expected to vary greatly depending on the soil parent material and other soil environmental factors. Detailed studies focused on clay mineral composition are also desirable.

The research group of the University of Tokyo and Forestry and Forest Products Research Institute, in cooperation with mushroom log or log-cultivated mushroom producers, started trial planting experiments of Konara oak or sawtooth oak *Quercus acutissima* (sawtooth oak), which was newly planted in an area where log production had been halted due to radioactive contamination. A treatment area was established with and without potassium fertilizer application. Although it will take another 12–15 years before the planted oak trees will be used as mushroom logs, it is important to keep in mind that the research should be conducted in dialogue with those producers to establish with the local people and effectively deliver the research results.

18.8 Conclusion

This chapter described the results of the studies the author has been conducting for the past 8 years to contribute to the resumption of hardwood forestry operations in radiation-contaminated mushroom log forests. The studies focused on using radiocesium activity concentrations in current-year shoots as an indicator to clarify the characteristics of radiocesium absorption from the soil by coppiced Konara oak to predict the future contamination levels. There has been significant progress in the studies of hardwood log forests. Although not mentioned in this chapter, important themes should be further addressed in the future, such as the effects of topography on radiocesium absorption by trees. It is hoped that conducting studies while keeping dialogue with residents and the forestry industry frontliners can contribute to the recovery of the livelihood of people in the affected areas and the forestry industry. In the field of forestry, which has a long production period, a shortcut to solving problems is to gain a deeper understanding of the scientific mechanisms by which Konara oak absorbs cesium from the soil and transports it within the tree. Further studies should focus on developing practical measures backed by scientific evidence.

Acknowledgements A series of studies was initiated after the author spent 2 years from 2013 at the Isotope Facility for Agricultural Education and Research of the Graduate School of Agricultural and Life Sciences of the University of Tokyo. The author would like to express deepest gratitude to Professors Tomoko Nakanishi, Keitaro Tanoi, and their laboratory staff and the researchers at the FFPRI who shared much of the research and analysis.

References

Eguchi T, Ohta T, Ishikawa T, Matsunami H, Takahashi Y, Kubo K, Yamaguchi N, Kihou N, Shinano T (2015) Influence of the nonexchangeable potassium of mica on radiocesium uptake by paddy rice. J Environ Radioact 147:33–42. https://doi.org/10.1016/j.jenvrad.2015.05.002

Hashimoto S (2023) Overview of radiocesium dynamics in forests: first decade and future perspectives. In: Nakanishi TM, Tanoi K (eds) Agricultural implications of the Fukushima nuclear accident (IV). Springer, Japan, Tokyo. https://doi.org/10.1007/978-981-19-9361-9_12

Hashimoto S, Matsuura T, Nanko K, Linkov I, Shaw G, Kaneko S (2013) Predicted spatio-temporal dynamics of radiocesium deposited onto forests following the Fukushima nuclear accident. Sci Rep 3:19526–19525. https://doi.org/10.1038/srep02564

Hashimoto S, Imamura N, Kaneko S, Komatsu M, Matsuura T, Nishina K, Ohashi S (2020) New predictions of 137Cs dynamics in forests after the Fukushima nuclear accident. Sci Rep 10:1–11. https://doi.org/10.1038/s41598-019-56800-5

Hashimoto S, Tanaka T, Komatsu M, Gonze M-A, Sakashita W, Kurikami H, Nishina K, Ota M, Ohashi S, Calmon P, Coppin F, Imamura N, Hayashi S, Hirai K, Hurtevent P, Koarashi J, Manaka T, Miura S, Shinomiya Y, Shaw G, Thiry Y (2021) Dynamics of radiocaesium within forests in Fukushima—results and analysis of a model inter-comparison. J Environ Radioact 238:106721. https://doi.org/10.1016/j.jenvrad.2021.106721

Hashimoto S, Komatsu M, Miura S (2022) Forest radioecology in Fukushima. Springer Nature, Cham. https://doi.org/10.1007/978-981-16-9404-2

Imamura N, Komatsu M, Ohashi S, Hashimoto S, Kajimoto T, Kaneko S, Takano T (2017) Temporal changes in the radiocesium distribution in forests over the 5 years after the Fukushima Daiichi Nuclear Power Plant accident. Sci Rep 7:1–11. https://doi.org/10.1038/s41598-017-08261-x

Imamura N, Watanabe M, Manaka T (2021) Estimation of the rate of 137Cs root uptake into stemwood of Japanese cedar using an isotopic approach. Sci Total Environ 755:142478. https://doi.org/10.1016/j.scitotenv.2020.142478

Kanasashi T, Miura S, Hirai K, Nagakura J, Itô H (2020) Relationship between the activity concentration of 137Cs in the growing shoots of *Quercus serrata* and soil 137Cs, exchangeable cations, and pH in Fukushima, Japan. J Environ Radioact 220:106276. https://doi.org/10.1016/j.jenvrad.2020.106276

Kato H, Onda Y, Gomi T (2012) Interception of the Fukushima reactor accident-derived 137Cs, 134Cs and 131I by coniferous forest canopies. Geophys Res Lett 39:2928. https://doi.org/10. 1029/2012GL052928

Kato N, Kihou N, Fujimura S, Ikeba M, Miyazaki N, Saito Y, Eguchi T, Itoh S (2015) Potassium fertilizer and other materials as countermeasures to reduce radiocesium levels in rice: results of urgent experiments in 2011 responding to the Fukushima Daiichi Nuclear Power Plant accident. Soil Sci Plant Nutr 61:179–190. https://doi.org/10.1080/00380768.2014.995584

Kenzo T, Saito S, Miura S, Kajimoto T, Kobayashi NI, Tanoi K (2020) Seasonal changes in radiocesium and potassium concentrations in current-year shoots of saplings of three tree species in Fukushima, Japan. J Environ Radioact 223:106409. https://doi.org/10.1016/j. jenvrad.2020.106409

Kobayashi NI, Kobayashi R (2023) Verification of uptake and transport properties of cesium in hydroponically cultivated *Quercus serrata*. In: Nakanishi TM, Tanoi K (eds) Agricultural implications of the Fukushima nuclear accident (IV). Springer, Japan, Tokyo. https://doi.org/ 10.1007/978-981-19-9361-9_5

Kobayashi R, Masumori M, Tange T, Tanoi K, Kobayashi NI, Miura S (2023) Effect of exchangeable and non-exchangeable potassium in soil on cesium uptake by *Quercus serrata* seedlings. In: Nakanishi TM, Tanoi K (eds) Agricultural implications of the Fukushima nuclear accident (IV). Springer, Japan, Tokyo. https://doi.org/10.1007/978-981-19-9361-9_15

Kobayashi NI, Ito R, Masumori M (2019a) Radiocesium contamination in forests and the current situation of growing oak trees for mushroom logs. In: Nakanishi TM, O'Brien M, Tanoi K (eds) Agricultural implications of the Fukushima nuclear accident (III): after 7 years. Springer, Singapore, pp 107–122. https://doi.org/10.1007/978-981-13-3218-0_11

Kobayashi R, Kobayashi NI, Tanoi K, Masumori M, Tange T (2019b) Potassium supply reduces cesium uptake in Konara oak not by an alteration of uptake mechanism, but by the uptake competition between the ions. J Environ Radioact 208:106032. https://doi.org/10.1016/j. jenvrad.2019.106032

Komatsu M, Kaneko S, Ohashi S, Kuroda K, Sano T, Ikeda S, Saito S, Kiyono Y, Tonosaki M, Miura S, Akama A, Kajimoto T, Takahashi M (2016) Characteristics of initial deposition and behavior of radiocesium in forest ecosystems of different locations and species affected by the Fukushima Daiichi nuclear power plant accident. J Environ Radioact 161:2–10. https://doi.org/ 10.1016/j.jenvrad.2015.09.016

Komatsu M, Hirai K, Nagakura J, Noguchi K (2017) Potassium fertilisation reduces radiocesium uptake by Japanese cypress seedlings grown in a stand contaminated by the Fukushima Daiichi nuclear accident. Sci Rep 7:15612. https://doi.org/10.1038/s41598-017-15401-w

Kuroda K, Kagawa A, Tonosaki M (2013) Radiocesium concentrations in the bark, sapwood and heartwood of three tree species collected at Fukushima forests half a year after the Fukushima Dai-ichi nuclear accident. J Environ Radioact 122:37–42. https://doi.org/10.1016/j.jenvrad. 2013.02.019

Kuroda K, Yamane K, Itoh Y (2020) Radial movement of minerals in the trunks of standing Japanese cedar (*Cryptomeria japonica* D. Don) trees in summer by tracer analysis. Forests 11: 562. https://doi.org/10.3390/f11050562

Masumori M, Kobayashi NI, Tanoi K, Nihei N, Miura S, Kanasashi T (2023) Challenge to resume production of mushroom bed logs by potassium fertilizer application. In: Nakanishi TM, Tanoi K (eds) Agricultural implications of the Fukushima nuclear accident (IV). Springer, Japan, Tokyo. https://doi.org/10.1007/978-981-19-9361-9_17

Miura S (2016) The effects of radioactive contamination on the forestry industry and commercial mushroom-log production in Fukushima, Japan. In: Nakanishi TM, Tanoi K (eds) Agricultural implications of the Fukushima nuclear accident. Springer Japan, Tokyo, pp 145–160. https:// doi.org/10.1007/978-4-431-55828-6_12

Rosén K, Vinichuk M, Nikolova I, Johanson K (2011) Long-term effects of single potassium fertilization on 137Cs levels in plants and fungi in a boreal forest ecosystem. J Environ Radioact 102:178–184. https://doi.org/10.1016/j.jenvrad.2010.11.009

Sakashita W, Miura S, Nagakura J, Kanasashi K (2023) Toward the estimation of radiocesium activity concentration in trunks of coppiced *Quercus serrata*: leaf availability in lieu of felling. In: Nakanishi TM, Tanoi K (eds) Agricultural implications of the Fukushima nuclear accident (IV). Springer, Japan, Tokyo. https://doi.org/10.1007/978-981-19-9361-9_13

Sakashita W, Miura S, Akama A, Ohashi S, Ikeda S, Saitoh T, Komatsu M, Shinomiya Y, Kaneko S (2020) Assessment of vertical radiocesium transfer in soil via roots. J Environ Radioact 222: 106369. https://doi.org/10.1016/j.jenvrad.2020.106369

Sakashita W, Miura S, Nagakura J, Kanasashi T, Shinomiya Y (2021) Seasonal stability of 137Cs in coppiced *Quercus serrata* current-year branches: toward the estimation of trunk 137Cs activity concentrations without felling. Ecol Indic 133:108361. https://doi.org/10.1016/j.ecolind.2021. 108361

Thiry Y, Albrecht A, Tanaka T (2018) Development and assessment of a simple ecological model (TRIPS) for forests contaminated by radiocesium fallout. J Environ Radioact 190:149–159. https://doi.org/10.1016/j.jenvrad.2018.05.009

Thiry Y, Tanaka T, Dvornik AA, Dvornik AM (2020) TRIPS 2.0: toward more comprehensive modeling of radiocaesium cycling in forest. J Environ Radioact 214:106171. https://doi.org/10. 1016/j.jenvrad.2020.106171

White PJ, Broadley MR (2000) Tansley review no. 113. Mechanisms of caesium uptake by plants. New Phytol 147:241–256. https://doi.org/10.1046/j.1469-8137.2000.00704.x

Zhu YG, Botany ESJ, of E (2000) Plant uptake of radiocaesium: a review of mechanisms, regulation and application. J Exp Bot 51(351):1635–1645. https://doi.org/10.1093/jexbot/51. 351.1635

Chapter 19
Contribution of Cesium-Bearing Microparticles to Cesium in Soil and River Water of the Takase River Watershed and Their Effect on the Distribution Coefficient

Takahiro Tatsuno, Hiromichi Waki, Waka Nagasawa, Naoto Nihei, Masashi Murakami, and Nobuhito Ohte

19.1 Introduction

Large amounts of radioactive materials were released due to the Fukushima Daiichi Nuclear Power Plant (FDNPP) accident on March 11, 2011 (Chino et al. 2011). In particular, radioactive cesium-137 (^{137}Cs) has a long half-life time of approximately 30 years, and thus, its effect on the environment continues for a long time in contaminated areas (Spezzano 2005). Because radioactive Cs (^{134}Cs and ^{137}Cs) is adsorbed onto clay minerals such as illite (Comans et al. 1991), a large amount of Cs remains in the surface soil layer (Koarashi et al. 2016; Takahashi et al. 2018). Therefore, decontamination activities such as the removal of vegetation and surface

The English in this document has been checked by at least two professional editors, both native speakers of English. For a certificate, please see: http://www.textcheck.com/certificate/ owKY9s.

T. Tatsuno (✉)
Institute of Environment Radioactivity, Fukushima University, Fukushima, Japan
e-mail: t.tatsuno@ier.fukushima-u.ac.jp

H. Waki · N. Ohte
Graduate School of Informatics, Kyoto University, Kyoto, Japan
e-mail: waki.hiromichi.54s@st.kyoto-u.ac.jp; nobu@i.kyoto-u.ac.jp

W. Nagasawa
Graduate School of Science and Engineering, Chiba University, Chiba, Japan

N. Nihei
Faculty of Food and Agricultural Science, Fukushima University, Fukushima, Japan
e-mail: nihei@agri.fukushima-u.ac.jp

M. Murakami
Graduate School of Science, Chiba University, Chiba, Japan
e-mail: muramasa@faculty.chiba-u.jp

soil have been undertaken in residential and agricultural areas since the accident (Evrard et al. 2020). On the other hand, as intensive decontamination activities have not been conducted in forest areas, which occupy approximately 70% of the area of Fukushima Prefecture, large amounts of radioactive materials most likely remain in forest soils (Takahashi et al. 2018; Yoschenko et al. 2022). Cs in the surface soil layer may enter the river due to soil erosion during rainfall (Evrard et al. 2015; Osawa et al. 2018; Niida et al. 2022). In other words, forests may be a source of Cs contamination in downstream areas. Residents have begun to return to heavily impacted areas such as Katsurao Village in 2022 and clarifying the subsequent remigration of radioactive materials from forests to these areas is essential.

Cs-bearing microparticles (CsMPs) were measured in soil and water samples collected in Fukushima Prefecture (Miura et al. 2018; Igarashi et al. 2019; Ikehara et al. 2020). CsMPs are insoluble glassy particles derived from the FDNPP (Adachi et al. 2013; Miura et al. 2018; Ikehara et al. 2020). They have ^{137}Cs concentration per unit mass more than 100,000 times higher than clay minerals that adsorbed ^{137}Cs (Igarashi et al. 2019). Because the particle size of CsMPs can be as small as a few micrometers, concern has been raised about the effects of internal exposure when CsMP is taken into the body of a living organism. Furthermore, as the Cs in CsMPs is contained in insoluble glassy particles, its release into the liquid phase of river water is more difficult than Cs adsorbed onto clay minerals (Miura et al. 2018; Okumura et al. 2020). The distribution coefficient (K_d) in rivers represents solute transition between the solid and liquid phases. Previous studies have suggested that when large amounts of particles that do not contribute to the Cs transition between the solid phase and the liquid phase are present, such as Cs in CsMPs, the apparent K_d may be higher than the net K_d associated with Cs transition (Konoplev et al. 2016; Miura et al. 2018). As Cs in the liquid phase (i.e., dissolved Cs) corresponds to bioavailable Cs, accurate evaluation of K_d is essential to understand the bioavailability of Cs to plants in surrounding ecosystems (Staunton 1994; Koarashi et al. 2016; Miura et al. 2021). On the other hand, the distribution of CsMPs in the forested watershed and the contributions of CsMPs to K_d and Cs concentrations in soils and river water remain unclear (Miura et al. 2021). Ikehara et al. (2020) investigated the distribution of CsMPs in soils of Fukushima Prefecture. They found a large amount of CsMPs in the soil near the FDNPP. However, the distribution of CsMP in their study was assessed immediately after the accident, and it is now necessary to determine the current distribution of CsMP 11 years after the accident. Furthermore, previous studies have not clarified the distribution of CsMPs in forest soil and its vertical distribution in soil to depths of several tens of centimeters. Assessing the migration of CsMPs from the forest to downstream areas and the effects of CsMP on river water requires comprehensive elucidation of the distribution of CsMPs in forest soils, which provide the source of CsMPs to the river, as well as the inflow of CsMPs into rivers.

In this study, we investigated the distribution of CsMPs in forest soil and the inflow of CsMPs to a river in a forested watershed in Fukushima Prefecture. We evaluated Cs derived from CsMPs as a proportion of the Cs concentrations in forest soil and river water.

19.2 Methods

19.2.1 Study Sites

We collected soil and river water samples from the Takase River watershed in Namie Town, Fukushima Prefecture, Japan. The study catchment (37°22′17.2″N to 37°32′37.2″N, 140°40′53.2″E to 140°56′44.4″E) has an area of 244 km^2 and the average Cs inventory (i.e., ^{137}Cs + ^{134}Cs) in the surface soil was 980 kBq m^{-2} in September 2017 (Fig. 19.1) (MEXT 2014). The collection sites of soil and water samples are located 10.8 km and 9.18 km northwest of the FDNPP, respectively.

19.2.2 Sampling and Analysis of Forest Soils

We collected soil samples from the sampling sites on September 5, 2021. Three soil column samples with soil depths of 0–20 cm were collected at 3 m intervals. After each soil sample was returned to the laboratory, the following treatment was conducted. The soil column samples were sliced into layers from the surface at intervals of one to several centimeters, and the soil sample from each depth was dried in an oven at 100 °C for 24 h. After drying, each sample was passed through a sieve with a pore size of 2.00 mm. The soil that passed through the sieve was placed into a 100 mL U8 container (Pla tsubo 3-20, Umano Kagaku Youki Co., Ltd., Osaka, Japan). The Cs concentration (sum of ^{134}Cs and ^{137}Cs) of the bulk soil sample was measured using a germanium semiconductor detector (GC4020, Canberra Industries Inc., USA).

After measurement of the Cs concentration, a portion of the soil sample (approximately 2–3 g dry soil) was placed into a zip-closure bag and Cs radioactivity derived from CsMPs was measured using an imaging plate (IP; BAS IP MS 2025, Cytiva,

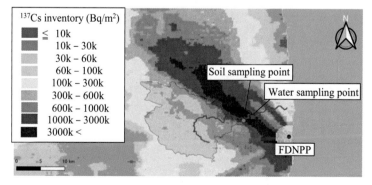

Fig. 19.1 Sampling sites for forest soils and river water in the Takase River watershed. ^{137}Cs inventory was calculated using data from the Ministry of Education, Culture, Sports, Science and Technology, Japan (MEXT 2014)

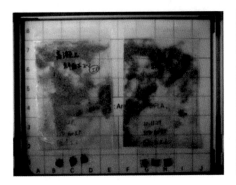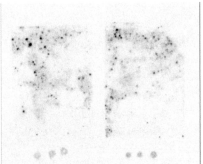

Fig. 19.2 Photograph and IP image of soil samples. Black dots indicate high Cs radioactivity

Japan) with an IP reader (FLA-9500, GE Healthcare Bio-Science AB, Sweden) (Sagawa et al. 2011; Miura et al. 2018; Ikehara et al. 2018) (Fig. 19.2). The IP can store energy emitted from radioactive materials and be read with an IP reader to create a luminescent image proportional to the radioactive energy. Cs radioactivity at each spot with high energy in the image was calculated using image processing software (ImageJ, National Institutes of Health, USA). In this study, spots with activity of 0.01 Bq or more were assumed to indicate Cs derived from CsMPs. Furthermore, we calculated the proportion of Cs concentration derived from CsMPs to the Cs concentration in the bulk soil at each soil depth, hereinafter referred to as the proportion of CsMPs in the soil with reference to the previous study (Ikehara et al. 2020).

19.2.3 Sampling and Analysis of River Water

River water samples were collected from the Takase River during rainfall events on June 29, July 7, and July 27, 2021, using an automatic programmable water sampler (6712, Teledyne ISCO, USA). Approximately 1.00 L river water was collected over a total of 8 h at 20 min intervals during one rainfall event (i.e., approximately 24 L sample was collected during one rainfall event).

The water samples were filtered in 1 L volumes using a membrane filter with a pore size of 0.45 μm (JHWP004700, Merck Co., Germany) and then divided into suspended solids (SS) and filtrate fractions. After drying the SS on the membrane filter in an oven at 100 °C for 24 h, it was placed in an U8 container, and its Cs concentration was measured using a germanium semiconductor detector. Cs in SS is referred to as particulate Cs. Next, Cs radioactivity derived from CsMP in the SS was measured using an IP as for the soil sample (Fig. 19.3). We determined the proportion of Cs concentration derived from CsMPs to the particulate Cs concentration for each SS sample, hereinafter called the proportion of CsMPs in particulate Cs.

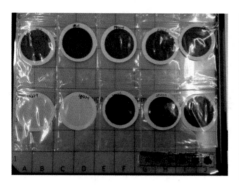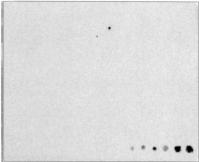

Fig. 19.3 Photograph and IP image of SS samples

For analysis of the filtrate, the whole sample from one rainfall event (i.e., approximately 24 L) was passed through a single cartridge filter (CS-13ZN, Japan Vilene Co., Japan), and the Cs concentration was measured. Cs in the filtrate is referred to as dissolved Cs. Then, we calculated K_d (i.e., apparent K_d) and K_d without CsMPs (i.e., net K_d) for each sample, respectively (Miura et al. 2018). The K_d was obtained by dividing particulate Cs concentration per unit mass of SS by dissolved Cs concentration in the river water. The K_d without CsMPs was calculated by calculated by subtracting the Cs concentration derived from CsMPs from the suspended Cs concentration and dividing it by the dissolved Cs concentration.

19.3 Results and Discussion

19.3.1 Vertical Distribution of CsMPs in Forest Soil

Figure 19.4 shows the vertical distribution of Cs concentration in each soil layer and in CsMPs, along with the proportion of CsMPs in the soil for each layer. CsMPs were detected in soil samples collected to a depth of 20 cm. This result indicated that CsMPs moved within the soil. Because Cs in the CsMPs is surrounded by an insoluble glassy particle (Adachi et al. 2013), the Cs may not react directly with soil. The Cs in CsMPs may not be fixed to the FES and therefore can move to deeper soil layers. The average proportion of CsMPs in the soil was less than 3% in all soil layers (Fig. 19.4). This result is consistent with previous studies demonstrating that the proportion of CsMPs is small, because many forms of Cs other than CsMPs were distributed near the nuclear power plant (Ikehara et al. 2020). Furthermore, as observed for Cs concentration in the soil, the Cs concentration derived from CsMPs and the proportion of CsMPs decreased with increasing soil depth. No significant difference in the proportion of CsMPs in the soil was found between the layer at 1–10 cm and the surface soil layer at soil depth of 0–1 cm. On the other hand, the proportion of CsMPs in the soil significantly decreased below the soil

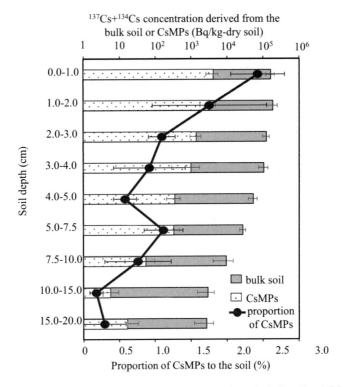

Fig. 19.4 Vertical distributions of Cs concentrations derived from the bulk soil and CsMPs, along with the proportion of CsMPs in the soil for each soil layer. Horizontal bars indicate the standard error calculated from three samples

depth of 10 cm ($p < 0.05$). These results suggest that CsMPs may migrate more slowly than ionic Cs in soil solution in the soil deeper layer. Few studies have investigated the vertical distribution and dynamics of CsMPs in soils. To elucidate the dynamics of CsMPs, further studies such as soil sampling and analysis of soil physics are needed in the future.

Figure 19.5 shows the variation in Cs concentration in the soil and Cs derived from CsMPs in each soil layer calculated from three soil column samples. In all layers, the Cs concentration derived from CsMPs had a greater coefficient of variation than Cs in the soil. This suggests that the distribution of CsMPs in soil might be more heterogeneous than the distribution of Cs adsorbed onto soil, and that these distributions might not always match.

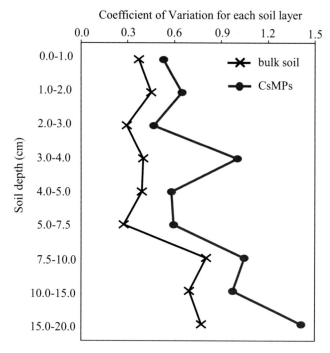

Fig. 19.5 Coefficients of variation for Cs concentration in the bulk soil and for Cs derived from CsMPs in each soil layer

19.3.2 Discharge of CsMPs from the Forested Catchment

Figure 19.6 shows changes in the particulate Cs concentration and Cs concentration derived from CsMPs for each sample. Figure 19.7 shows relationships of SS concentrations with particulate Cs concentrations and Cs concentrations derived from CsMPs. Particulate Cs was detected in all water samples (Fig. 19.6), and its concentration was positively correlated with SS concentrations (correlation coefficient, $r = 0.398$). This is consistent with previous studies showing that the discharge of particulate Cs increases with sediment runoff via soil erosion (Evrard et al. 2015; Osawa et al. 2018; Niida et al. 2022). On the other hand, CsMPs were not detected in some water samples (Fig. 19.6). Furthermore, no correlation was found between the Cs concentrations derived from CsMPs and SS concentrations ($r = 0.070$ in Fig. 19.7b). As noted in Sect. 19.3.1, Cs derived from CsMPs may be more unevenly distributed in the soil than Cs adsorbed onto the soil (Fig. 19.5). Therefore, the influx of CsMPs into rivers due to soil erosion may be less stable than the input of particulate Cs. However, soil sampling in this study was conducted at only one site in the watershed, and therefore, the tendency of CsMPs in the entire watershed cannot be addressed. Determining the distribution of CsMPs over a wide area, such as with a Cs inventory converted from air dose rates by aircraft monitoring (MEXT

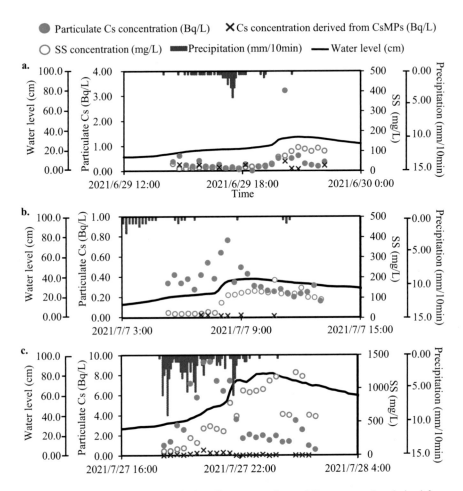

Fig. 19.6 Temporal changes in particulate Cs concentration and Cs concentration derived from CsMPs

2014), is technically difficult. Therefore, clarifying the relationship between the inflow of CsMPs into rivers and the spatial distribution of CsMPs in future research will require surveys in relatively narrow watersheds.

Table 19.1 shows the average Cs concentration by form (i.e., particulate Cs, dissolved Cs, and CsMPs), the proportion of CsMPs in particulate Cs, and the distribution coefficient for each sample. Particulate Cs accounted for more than 90% of Cs discharge during rainfall, in accordance with previous studies (Osawa et al. 2018; Niida et al. 2022). On the other hand, the average proportion of CsMPs in particulate Cs was only 3.46%, indicating that most of the discharged particulate Cs was in forms other than CsMPs. Furthermore, no significant difference was found between K_d and K_d without CsMPs for any sample ($p > 0.05$). Thus, CsMPs may

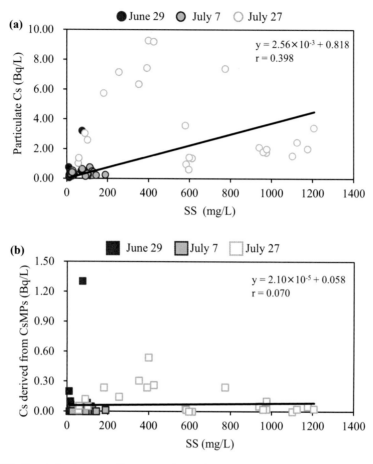

Fig. 19.7 Correlation between SS concentration and (**a**) particulate Cs concentration and (**b**) Cs concentration derived from CsMPs

not significantly affect the K_d of rivers, because the amount of Cs derived from CsMPs is smaller than the amount of Cs present in other forms.

19.4 Conclusion

CsMPs were detected from forest soils and river water in the Takase River watershed in 2021. The proportion of Cs concentration derived from CsMPs in the forest soils was as small as about 3%; therefore, they did not significantly affect the particulate Cs and distribution coefficients in the river water collected during rainfall. It was suggested that the variation of Cs concentration derived from CsMPs in the forest soil was greater than that in the bulk soil and affects the instability of CsMPs flowing

Table 19.1 Average Cs concentration, proportion of CsMPs in particulate Cs, and distribution coefficients for each sample

Sampling date	SS conc. (mg L^{-1})	Cs conc. in river water (Bq L^{-1})				Cs conc. per unit mass of SS (Bq kg^{-1})			Proportion of CsMPs (%)	K_d	
		Particulate	CsMPs	Particulate without CsMPs	Dissolved	Particulate	Derived from CsMP	Particulate without CsMPs			Without CsMPs
29-Jun-21	46.5 (±8.12)	0.418 (±0.127)	0.078 (±0.053)	0.339 (±0.075)	0.027	1.26 × 10^4 (±0.294 × 10^4)	0.195 × 10^4 (±0.099 × 10^4)	1.07 × 10^4 (±0.205 × 10^4)	7.11 (±2.38)	4.64 × 10^5 (±1.08 × 10^5)	3.92 × 10^5 (±0.76 × 10^5)
7-Jul-21	89.7 (±10.4)	0.360 (±0.028)	0.004 (±0.002)	0.355 (±0.028)	0.021	0.731 × 10^4 (±0.134 × 10^4)	0.009 × 10^4 (±0.005 × 10^4)	0.722 × 10^4 (±0.132 × 10^4)	0.95 (±0.37)	3.50 × 10^5 (±0.64 × 10^5)	3.46 × 10^5 (±0.63 × 10^5)
27-Jul-21	606 (±77.9)	3.57 (±0.564)	0.108 (±0.026)	3.465 (±0.540)	0.042	1.17 × 10^4 (±0.233 × 10^4)	0.037 × 10^4 (±0.009 × 10^4)	1.13 × 10^4 (±0.224 × 10^4)	2.33 (±0.34)	2.77 × 10^5 (±0.55 × 10^5)	2.68 × 10^5 (±0.53 × 10^5)
Average	2047 (±39.8)	1.450 (±0.261)	0.063 (±0.020)	1.390 (±0.250)	0.030 (±0.006)	1.05 × 10^4 (±0.134 × 10^4)	0.081 × 10^4 (±0.034 × 10^4)	0.973 × 10^4 (±0.111 × 10^4)	3.46 (±0.86)	3.64 × 10^5 (±0.46 × 10^5)	3.35 × 10^5 (±0.37 × 10^5)

into rivers. To grasp the dynamics of CsMPs via soil erosion, further research is needed on the dynamics of CsMPs in soils before they flow into rivers and their distribution in soil in detail.

Acknowledgements This work was supported by the Japan Society for the Promotion of Science, grant 20H00435, the FY2021 Research Fund of Environmental Radioactivity Research Network (F-21-28), and the academic fund of the Japanese Society of Irrigation, Drainage and Rural Engineering. We also acknowledge support from the Japan Atomic Energy Agency.

References

Adachi K, Kajino M, Zaizen Y, Igarashi Y (2013) Emission of spherical cesium-bearing particles from an early stage of the Fukushima nuclear accident. Sci Rep 3:12–15. https://doi.org/10.1038/srep02554

Chino M, Nakayama H, Nagai H, Terada H, Katata G, Yamazawa H (2011) Preliminary estimation of release amounts of 131 I and 137 Cs accidentally discharged from the Fukushima Daiichi Nuclear Power Plant into the atmosphere. J Nucl Sci Technol 48:1129–1134. https://doi.org/10.1080/18811248.2011.9711799

Comans RNJ, Hller M, De Preter P (1991) Sorption of Cs on illite nonequilibrium behavior and reversibility. J Phys Act Health 55:433–440. https://doi.org/10.4319/lo.2013.58.2.0489

Evrard O, Laceby JP, Lepage H, Onda Y, Cerdan O, Ayrault S (2015) Radiocesium transfer from hillslopes to the Pacific Ocean after the Fukushima nuclear power plant accident: a review. J Environ Radioact 148:92–110. https://doi.org/10.1016/j.jenvrad.2015.06.018

Evrard O, Laceby JP, Nakao A (2020) Effectiveness of landscape decontamination following the Fukushima nuclear accident: a review

Igarashi Y, Kogure T, Kurihara Y, Miura H, Okumura T, Satou Y, Takahashi Y, Yamaguchi N (2019) A review of Cs-bearing microparticles in the environment emitted by the Fukushima Dai-ichi nuclear power plant accident. J Environ Radioact 205:101–118. https://doi.org/10.1016/j.jenvrad.2019.04.011

Ikehara R, Suetake M, Komiya T, Furuki G, Ochiai A, Yamasaki S, Bower WR, Law GTW, Ohnuki T, Grambow B, Ewing RC, Utsunomiya S (2018) Novel method of quantifying radioactive cesium-rich microparticles (CsMPs) in the environment from the Fukushima Daiichi nuclear power plant. Environ Sci Technol 52:6390–6398. https://doi.org/10.1021/acs.est.7b06693

Ikehara R, Morooka K, Suetake M, Komiya T, Kurihara E, Takehara M, Takami R, Kino C, Horie K, Takehara M, Yamasaki S, Ohnuki T, Law GTW, Bower W, Grambow B, Ewing RC, Utsunomiya S (2020) Abundance and distribution of radioactive cesium-rich microparticles released from the Fukushima Daiichi nuclear power plant into the environment. Chemosphere 241:125019. https://doi.org/10.1016/j.chemosphere.2019.125019

Koarashi J, Nishimura S, Nakanishi T, Atarashi-Andoh M, Takeuchi E, Muto K (2016) Post-deposition early-phase migration and retention behavior of radiocesium in a litter-mineral soil system in a Japanese deciduous forest affected by the Fukushima nuclear accident. Chemosphere 165:335–341. https://doi.org/10.1016/j.chemosphere.2016.09.043

Konoplev A, Golosov V, Laptev G, Nanba K, Onda Y, Takase T, Wakiyama Y, Yoshimura K (2016) Behavior of accidentally released radiocesium in soil-water environment: looking at Fukushima from a Chernobyl perspective. J Environ Radioact 151:568–578. https://doi.org/10.1016/j.jenvrad.2015.06.019

Ministry of Education Culture, Sports and T (2014) Results of deposition of radioactive cesium of the airborne monitoring survey in the areas to which evacuation orders have been issued. https://radioactivity.nsr.go.jp/ja/contents/7000/6749/24/191_258_0301_18.pdf. Accessed 1 May 2022

Miura H, Kurihara Y, Sakaguchi A, Tanaka K, Yamaguchi N, Higaki S, Takahashi Y (2018) Discovery of radiocesium-bearing microparticles in river water and their influence on the solid-water distribution coefficient (K_d) of radiocesium in the Kuchibuto River in Fukushima. Geochem J 52:145–154. https://doi.org/10.2343/geochemj.2.0517

Miura H, Kurihara Y, Takahashi Y (2021) Environmental dynamics of radiocesium-bearing microparticles emitted from the Fukushima Daiichi nuclear power plant accident: transport from rivers to the ocean and impact on the environment. Geochemistry 55:122–131. https://doi.org/10.14934/chikyukagaku.55.122

Niida T, Wakiyama Y, Takata H, Taniguchi K, Kurosawa H, Fujita K, Konoplev A (2022) A comparative study of riverine 137Cs dynamics during high-flow events at three contaminated river catchments in Fukushima. Sci Total Environ 821:153408. https://doi.org/10.1016/j.scitotenv.2022.153408

Okumura T, Yamaguchi N, Kogure T (2020) Distinction between radiocesium (RCs)-bearing microparticles and RCs-sorbing minerals derived from the Fukushima nuclear accident using acid treatment. Chem Lett 49:1294–1297. https://doi.org/10.1246/CL.200374

Osawa K, Nonaka Y, Nishimura T, Tanoi K, Matsui H, Mizogichi M, Tatsuno T (2018) Quantification of dissolved and particulate radiocesium fluxes in two rivers draining the main radioactive pollution plume in Fukushima, Japan (2013–2016). Anthropocene 22:40–50. https://doi.org/10.1016/j.ancene.2018.04.003

Sagawa N, Yamazaki T, Kurosawa S, Izaki K (2011) JAEA-technology application of imaging plate to the radiation protection in the MOX fuel fabrication facility march 2011 Japan Atomic Energy Agency. IAEA, Vienna

Spezzano P (2005) Distribution of pre- and post-Chernobyl radiocaesium with particle size fractions of soils. J Environ Radioact 83:117–127. https://doi.org/10.1016/j.jenvrad.2005.02.002

Staunton S (1994) Adsorption of radiocaesium on various soils: consequences of the effects of soil: solution composition on the distribution coefficient. Eur J Soil Sci 45:409–418. https://doi.org/10.1111/j.1365-2389.1994.tb00526.x

Takahashi J, Onda Y, Hihara D, Tamura K (2018) Six-year monitoring of the vertical distribution of radiocesium in three forest soils after the Fukushima Dai-ichi nuclear power plant accident. J Environ Radioact 192:15. https://doi.org/10.1016/j.jenvrad.2018.06.015

Yoschenko V, Nanba K, Wada T, Johnson TE, Zhang J, Workman D, Nagata H (2022) Late phase radiocesium dynamics in Fukushima forests post deposition. J Environ Radioact 251:106947. https://doi.org/10.1016/j.jenvrad.2022.106947

Chapter 20
Global Fallout: Radioactive Materials from Atmospheric Nuclear Tests That Fell Half a Century Ago and Where to Find Them

Eriko Ito, Satoru Miura, Michio Aoyama, and Koji Shichi

20.1 Global Fallout

From the first atmospheric nuclear weapon test at Alamogordo, New Mexico, conducted by the USA in July 1945, to the last atmospheric weapon test at Lop Nor conducted by the People's Republic of China in October 1980, several countries have conducted a series of nuclear weapons tests in the atmosphere. The large quantities of anthropogenic radioactive nuclides injected into the stratosphere originated from nuclear weapons tests that gradually migrated to the troposphere (0.5–1.7 years, Hirose et al. 1987) and fell as radioactive fallout on land and ocean surfaces throughout the globe, so-called global fallout. The total amount of radiation released from atmospheric nuclear weapons tests peaked in the 1950s and early 1960s. Global fallout included various radionuclides such as ^{90}Sr, ^{137}Cs, ^{239}Pu, and ^{240}Pu. Among them, ^{137}Cs originated from the global fallout (hereafter, ^{137}Cs-GFO) and has attracted significant social attention for its environmental impact due to its long half-life (30.2 years) and has been used in studies of soil erosion processes. Two

E. Ito (✉)
Hokkaido Research Center, Forestry and Forest Products Research Institute, Sapporo, Japan
e-mail: iter@ffpri.affrc.go.jp

S. Miura
Forestry and Forest Products Research Institute, Tsukuba, Japan
e-mail: miura@ffpri.affrc.go.jp

M. Aoyama
Faculty of Life and Environmental Sciences, Center for Research in Isotopes and Environmental Dynamics, University of Tsukuba, Tsukuba, Japan
e-mail: michio.aoyama@ied.tsukuba.ac.jp

K. Shichi
Shikoku Research Center, Forestry and Forest Products Research Institute, Kochi, Japan
e-mail: shichi@ffpri.affrc.go.jp

© The Author(s) 2023
T. M. Nakanishi, K. Tanoi (eds.), *Agricultural Implications of Fukushima Nuclear Accident (IV)*, https://doi.org/10.1007/978-981-19-9361-9_20

regions of high ^{137}Cs fallout characterize the global spatial distribution of ^{137}Cs fallout; these areas are located on the eastern sides of the Eurasian continent and the North American continent in the Northern Hemisphere. The Japanese archipelago is in the high deposition region (Aoyama 2018); therefore, it had been exposed to radioactive contamination even before the Chernobyl and Fukushima nuclear accidents. We address the consequences of ^{137}Cs-GFO over the Japanese archipelago.

This chapter was based on Ito et al. (2020), but with an additional discussion on the vertical distribution of ^{137}Cs-GFO in forest soils and derived sediment migration on forested hillslopes over 50 years.

20.2 Cumulative Deposition of ^{137}Cs-GFO (From Monthly Data Collected by Meteorological Observatories and Local Authorities)

Many stations across Japan have observed the ^{137}Cs-GFO deposits since the late 1950s. The monthly deposit of ^{137}Cs-GFO observed by the Meteorological Research Institute (MRI, located in Tokyo until March 1980 and at Tsukuba after April 1980) and its decay-corrected cumulative deposit was shown as time-series data (Fig. 20.1). The monthly deposit of ^{137}Cs-GFO reached its maximum in the early 1960s (0.55 kBq m^{-2} month^{-1}, June 1963). The decay-corrected cumulative ^{137}Cs-GFO peaked in the late 1960s, with a maximum value of ~6.05 kBq m^{-2} (June–August 1966). This was 55–75% of the monthly ^{137}Cs deposit in Tokyo immediately after the Fukushima Daiichi Nuclear Power Plant (FDNPP) accident (8.1–11.0 kBq m^{-2}, March 2011, Shinjuku, Tokyo).

The decay-corrected cumulative deposit of ^{137}Cs-GFO as of October 1, 2008 (hereafter, referred to as cumulative fallout), was estimated by adding up the decay-corrected monthly deposit of ^{137}Cs-GFO (including reconstructed values before the start of ^{137}Cs observation from July 1945 to March 1957, see also footnote in Fig. 20.1 for data source). The cumulative fallout in Tokyo from July 1945 to September 2008 was estimated to be 2.73 kBq m^{-2} as of October 1, 2008 (Ito et al. 2020), and the average elapsed time from ^{137}Cs-GFO fallout, as of October 1, 2008, was 45.2 years. Thus, the theoretical value of the cumulative fallout was ~45% of the maximum in the late 1960s. This value for Tokyo is the most accurate estimate of the cumulative fallout in Japan, since the measurement in Tokyo was the most prolonged and uninterrupted in Japan. However, regional differences in global fallout within Japan were substantial, as will be shown later, and the estimated value for Tokyo does not necessarily give a representative cumulative fallout value for Japan.

The cumulative fallout was reconstructed for 39 stations that started observations before 1980 (including MRI-Tsukuba, which started its observation in April 1980). The amount of ^{137}Cs-GFO deposition during the no-observation period was determined by proportional calculation using other stations' values as reference. Details

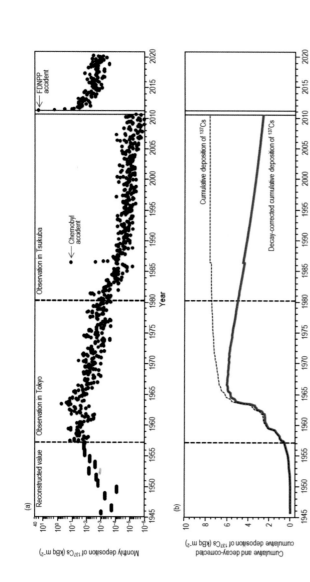

Data source: IGFD database, Aoyama 2019; Environmental Radiation Database, the Secretariat of the Nuclear Regulation Authority, https://search.kankyo-hoshano.go.jp/servlet/search.top

Fig. 20.1 Deposition of ^{137}Cs observed by the Meteorological Research Institute. (**a**) Monthly ^{137}Cs deposition from July 1945 to March 2020. (**b**) Cumulative (thin black dashed line) and decay-corrected cumulative (thick gray line) deposition of ^{137}Cs from July 1945 to July 2010. Data source: IGFD database, Aoyama (2019); Environmental Radiation Database, the Secretariat of the Nuclear Regulation Authority, https://search.kankyo-hoshano.go.jp/servlet/search.top. The

of the reconstruction method are shown elsewhere (Ito et al. 2020). The average ($\pm$SD) cumulative fallout was 2.47 $\pm$ 0.95 kBq m^{-2} ($n = 39$), ranging from 1.23 to 5.00 kBq m^{-2} (Fig. 20.2a, revised from Ito et al. 2020). Figure 20.3 shows the spatial distribution of the reconstructed cumulative fallout (revised from Ito et al. 2020). It was more abundant on the Sea of Japan side of northern Japan (e.g., Akita and Fukui) and less abundant on the Pacific side of western Japan and the Seto Inland Sea coast (e.g., Osaka and Okayama). A numerical list of the reconstructed values has been provided elsewhere (Ito et al. 2020).

The ^{137}Cs emitted from the Chernobyl nuclear power plant accident (April 26, 1986) also accumulated in the forest soil in Japan before the FDNPP accident. Most of the Chernobyl-generated ^{137}Cs fallout at the MRI (Tsukuba, Japan) occurred during May 1986 (Aoyama et al. 1986, 1987; Higuchi et al. 1988, Fig. 20.1). The amount of ^{137}Cs fallout from May 1986 to April 1987 (within a year after the Chernobyl accident) and the amount of ^{137}Cs fallout from May 1986 to September 2008 correspond to less than 3% and 5%, respectively, of the decay-corrected cumulative fallout as of October 2008 (Fig. 20.1). Thus, most of the ^{137}Cs detected in Japanese forests before the FDNPP accident came from the global fallout.

20.3   Residual ^{137}Cs-GFO in Forest Soils in Japan Before the FDNPP Accident

The FDNPP accident (March 11, 2011) resulted in an additional 2.7 PBq of ^{137}Cs deposition onto the terrestrial environment in Japan (Onda et al. 2020). Airborne monitoring survey reported that ^{137}Cs deposit on the ground surface within the zone 80 km from the FDNPP ranged from 10 to >3000 kBq m^{-2} (Nuclear Regulation Authority, https://radioactivity.nsr.go.jp/en/list/307/list-1.html). This value ranged from 1.3 to >400 times higher than the simple (i.e., nondecay-corrected) cumulative amount of ^{137}Cs-GFO in Tokyo mentioned above. The majority (approximately 70%) of the areas where the FDNPP-generated radioactive materials fell were forests. Among the contaminated radioactive cesium (^{137}Cs and ^{134}Cs), the consequence of ^{137}Cs, which has a considerably long half-life of 30.2 years, in forested areas has been a significant concern for local inhabitants (Onda et al. 2020).

Many studies on the redistribution of ^{137}Cs fallout, both ^{137}Cs-GFO (Jagercikova et al. 2015) and the Chernobyl-generated ^{137}Cs (Rafferty et al. 2000; Zhiyanski et al. 2008), have revealed the long-term persistence of ^{137}Cs in soil upon its adsorption

---

Fig. 20.1 (continued) MRI was located in Tokyo until March 1985 and at Tsukuba after April 1985. From July 1945 to March 1957, before the start of ^{137}Cs observation, data were interpolated with reconstructed values (Katsuragi 1983; Aoyama 1999). From August 2010 to March 2011, the monthly ^{137}Cs observation was missing due to contamination of the experimental and measurement environment caused by the FDNPP accident

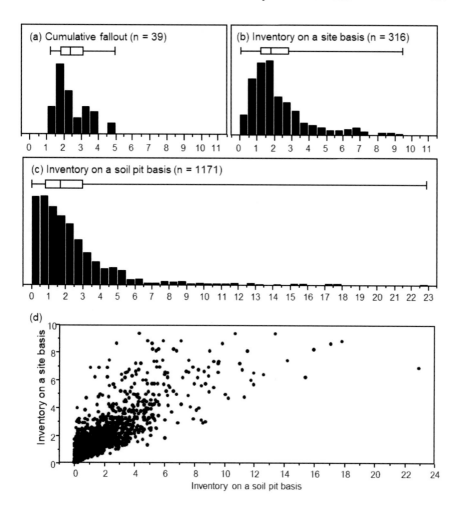

Fig. 20.2 Global fallout of radiocesium 137 (^{137}Cs-GFO) in Japan (kBq m^{-2}). (**a**) Frequency distribution of cumulative fallout: the decay-corrected cumulative depositions of ^{137}Cs-GFO for fallout observatories. (**b**) Frequency distribution of inventory data on a plot basis: simple arithmetic mean of ≤4 soil pits in the NFSCI survey site of the ^{137}Cs-GFO inventory in 0–30 cm deep forest soils. (**c**) Frequency distribution of inventory data on a soil pit basis: the ^{137}Cs-GFO inventory in 0–30 deep forest soils for each soil pit. (**d**) Relations of the ^{137}Cs-GFO inventory on a site basis against the ^{137}Cs-GFO inventory on a soil pit basis. The value of ^{137}Cs-GFO is shown as of October 1, 2008. The box–whisker plots mark the minimum and maximum and the 25th, 50th, and 75th percentile points. [Figures (**a–c**) were revised from Ito et al. (2020)]

onto clays (Dyer et al. 2000). The FDNPP-generated ^{137}Cs has also been expected and demonstrated to remain in the surface layer of the forest soil and decrease in the long-term due to radioactive decay. This prospect has been substantiated, although it was based on a fewer than 10-year study period (Onda et al. 2020). However, Japan is prone to sediment-related disasters (e.g., mass movements and soil erosion) due to

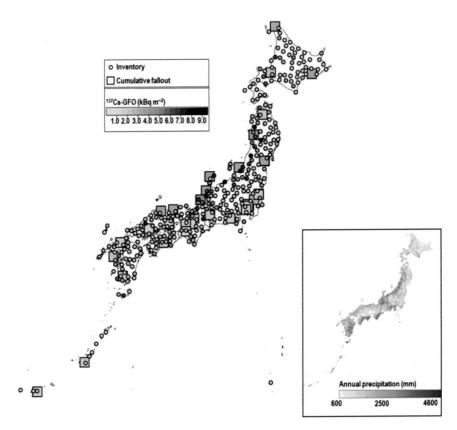

Fig. 20.3 Spatial distribution of ^{137}Cs-GFO and precipitation across Japan. Forest soil inventory on a plot basis (circle, 0–30 cm depth) and cumulative fallout (square) (kBq m^{-2}). The value of ^{137}Cs-GFO is shown as of October 1, 2008. Normal total annual precipitation was also shown in the lower right (mm). [Revised from Ito et al. (2020)]

a combination of various geographical factors (e.g., rainfall, snowmelt, earthquake, soil, geological structure, and topography) (Yoshimatsu and Abe 2006). Such disasters may result in a specific redistribution pattern of ^{137}Cs (Fukuyama et al. 2005), which could lead to discharge from the forest watershed. Some of the FDNPP-generated ^{137}Cs has been reported to be discharged from forests into rivers, owing to sediment runoff caused by the heavy rainfall and steep terrain (Ueda et al. 2013; Shinomiya et al. 2014; Tsuji et al. 2016; Iwagami et al. 2017). Although FDNPP-contaminated forest areas continue to represent more stable contaminant stores (Taniguchi et al. 2019), the residual status of ^{137}Cs in Japan after a long period could be different from that of Chernobyl.

Therefore, we investigated the extent to which ^{137}Cs-GFO, which fell into forests about half a century ago (around 1960), remained in forest soil in the late 2000s. This is an attempt to clarify the long-term persistence of ^{137}Cs in Japan and to verify whether the prediction of the behavior of ^{137}Cs based on the Chernobyl study (^{137}Cs is retained in the surface layer of forest soil for a long time) is also applicable to the Japanese archipelago.

The Forestry and Forest Products Research Institute (FFPRI) has been storing soil samples from forests throughout Japan collected just before the FDNPP accident [the National Forest Soil Carbon Inventory (NFSCI) project, 2006–2010] (Ugawa et al. 2012; Nanko et al. 2017). We used the NFSCI soil samples to clarify how much ^{137}Cs had accumulated in Japanese forests before the FDNPP accident (Ito et al. 2020). As mentioned in the previous section, this ^{137}Cs was mainly a global fallout, with some (~3%) from the Chernobyl nuclear accident.

The ^{137}Cs inventory in the surface layer of mineral soil (30 cm) at the forests was estimated at 316 spatially uniformly selected locations from the NFSCI survey sites across Japan. Repeated subsamplings with four soil pits were conducted at each site. The locations of the soil pits were systematically determined, where the four soil pits were made at four directions (N, E, W, and S) on a circle of 35.68 m in diameter centered on the NFSCI survey site reference point located on the latitude–longitude grid (Ugawa et al. 2012).

Mineral soil samples in each soil pit were collected from three layers: 0–5, 5–15, and 15–30 cm. From the ^{137}Cs concentration in the soil samples and the soil bulk density, the inventory of ^{137}Cs per square meter was estimated up to the surface soil layer of 30 cm. The activity concentration of ^{137}Cs measurement was performed using an NaI well-type scintillation counter (2480 WIZARD2 Automatic Gamma Counter; PerkinElmer, Inc., Waltham, MA, USA). This NaI gamma counter system can perform measurements with an expanded uncertainty of 6.6% for certified materials. Details of the calibration, certification, and detection limit are shown elsewhere (Ito et al. 2020). The ^{137}Cs-GFO inventory was calculated by multiplying the ^{137}Cs activity concentration, which was decay-corrected to October 1, 2008 (i.e., the intermediate point in the soil-sampling period) data using a half-life of 30.2 years for ^{137}Cs, a layer thickness of 30 cm, and the bulk density of the soil sample.

The ^{137}Cs-GFO inventory in 0–30 cm deep forest soils (hereafter referred to as inventory) was estimated on-site. The average ± SD (±SE) of 316 sampling sites was 2.27 ± 1.73 (±0.10) kBq m^{-2} (ranging from 0.09 to 9.43 kBq m^{-2}, $n = 316$) (Ito et al. 2020). A right-skewed distribution, peaking at ~1.5 kBq m^{-2}, was observed (Fig. 20.2b). This skewness was somewhat similar to that of the cumulative fallout (Fig. 20.2a), but the distribution differed, because there were several values <1.0 kBq m^{-2} and a long-tail distribution in the range above 6.0 kBq m^{-2}. A wider range and an L-shaped frequent distribution were presented on a soil pit basis, and the average ± SD (±SE) of 1171 soil pits was 2.29 ± 2.30 (±0.07) kBq m^{-2} (ranging from 0.00 to 22.89 kBq m^{-2}, Fig. 20.2c). The more right-skewed distribution indicated a high frequency of sites (or pits) with inventories below the average and a low frequency but still a high number of sites (or pits) with values well above the average. Figure 20.2d shows the soil pit inventory versus the on-site inventory (i.e., an average of the inventory on a soil pit basis). A large variation in the inventory could be observed among the ~4 soil pits even within the same site. The forest redistribution of ^{137}Cs-GFO implied by Fig. 20.2d will be discussed in detail later in Sect. 20.5.

The inventory was also characterized by its spatial distribution. The Pacific coast had low inventory, and the Japanese coast of northern Japan had high inventory (Fig. 20.3, revised from Ito et al. 2020). Comparing this with the spatial distribution of the total annual precipitation (lower right panel, Fig. 20.3), we found that the inventory did not precisely correspond to the total annual precipitation. The low correlation between fallout and precipitation amount is characteristic of Japan (Malakhov and Pudovkina 1970). Our study also confirmed a previous finding (Ito et al. 2020).

The air dose from ^{137}Cs-GFO was sufficiently low compared to that from natural radiation sources. The monthly external exposure was 0.0021 (for adults) and 0.0027 mSv (for newborns) (Omori et al. 2020), corresponding to an average inventory value of ^{137}Cs-GFO (2.27 kBq m^{-2}) on October 1, 2008. Note that dose conversion coefficients for external exposure to ^{137}Cs distributed in the soil of 1.28E−06 (adults) and 1.67E−06 mSv/h per kBq m^{-2} (newborns) were used in this calculation (Satoh et al. 2014), and it was assumed that the ^{137}Cs radiation source was distributed as 0.5 g cm^{-2} at the soil surface. This is an extremely low value compared to the 0.07 mSv external exposure from cosmic and terrestrial radiation and a similar low value of 0.2 mSv/month from all natural radiation sources (UNSCEAR 2020).

20.4 Does ^{137}Cs-GFO Remain in the Forest's Surface Soil After 50 Years?

The average (±SD) cumulative ^{137}Cs-GFO for fallout observatories was 2.47 (±0.95) kBq m^{-2} ($n = 39$) (see Sect. 20.2), whereas the average (±SD) inventory of ^{137}Cs-GFO in forest soils on a sampling site basis was 2.27 (±1.73) kBq m^{-2} ($n = 316$) (see Sect. 20.3), as of October 1, 2008 (Ito et al. 2020). Statistically testing whether these values were the same is equivalent to determining whether ^{137}Cs-GFO remained in the forest after ~50 years. The inventory (i.e., the inventory of ^{137}Cs-GFO in 0–30-cm deep forest soils) and the cumulative fallout (i.e., the decay-corrected cumulative deposits of ^{137}Cs-GFO for fallout observatories) were both corrected to the values, as of October 1, 2008. However, we determined that it would be inappropriate to compare inventory and cumulative fallout on that basis alone. The locations of meteorological stations and forest soil research sites did not coincide (Fig. 20.3). The forest soil survey sites were evenly spaced throughout Japan, but the spatial distribution of available meteorological stations was not uniform. To properly compare the inventory and cumulative fallout, which were data sets collected at different locations, the statistical analysis was processed to adjust for total annual precipitation and regional differences.

We used a general linear mixed-effect (GLM) model to estimate ^{137}Cs-GFO using total annual precipitation, regions, and dataset categories (i.e., inventory or cumulative fallout) as fixed effects and sampling sites for NFSCI as a random effect.

Post hoc comparisons between the dataset categories were examined using the Tukey-HSD test ($p < 0.05$). In other words, the GLM model determined the inventory's least-squares means (LSM), and the cumulative fallout normalized the effects of total annual precipitation and region and tested whether there was a statistically significant difference between the inventory and the cumulative fallout. Details of the GLM analysis are shown elsewhere (Ito et al. 2020).

Note that the NFSCI forest soil survey sites were incorporated as a random effect in the model. This allowed us to adequately address the soil survey's sampling design, which has four repeated soil pits at each NFSCI site, into the statistical analysis. Eventually, this random effect incorporation enabled us to reveal the ^{137}Cs-GFO redistribution via ^{137}Cs-GFO-contaminated sediment migration on the forested hillslopes.

The GLM analysis indicated that the inventory and the cumulative fallout were not significantly different ($p = 0.34$, Ito et al. 2020). It was empirically demonstrated that most of the ^{137}Cs-GFO remained on the soil surface even half a century after the fallout. However, the inventory's model-estimated LSM (2.08 kBq m^{-2}) was maintained as 79.1% of the cumulative fallout (2.63 kBq m^{-2}) (Ito et al. 2020). These results will be discussed in more detail at the end of the next section.

20.5   Vertical Distribution of ^{137}Cs-GFO Reveals Sediment Redistribution Over 50 Years

20.5.1   Downward Infiltration in Soils Alone Cannot Explain the Various Vertical Distribution Patterns of ^{137}Cs-GFO

The inventory varied considerably among the four soil pits in the same NFSCI site (see Fig. 20.2d and Sect. 20.3). This was demonstrated in a GLM analysis using the ^{137}Cs-GFO inventory as an objective variable and the NFSCI sampling plots as a random effect term, where the ratio of the variance components of the site as a random effect to the total variance was estimated. It demonstrated that the variance component estimate within the site variability (58.8%) was greater than the variability among the site (41.2%). Why was the difference in ^{137}Cs-GFO inventory within sites (i.e., between locations less than 30 m away) greater than the difference among sites (i.e., between locations at least 30 km away)? We thought that it would be implausible to explain this by the spatial variation in the amounts of ^{137}Cs-GFO deposits alone.

We analyzed the soil's vertical distribution pattern of ^{137}Cs-GFO to determine the reason for this intra-site variability in the ^{137}Cs-GFO inventory. As noted in Sect. 20.3, the soil samples used in this study were collected from three layers of each soil pit: 0–5, 5–15, and 15–30 cm. A comprehensive meta-analysis estimated that the ^{137}Cs penetration velocities ranged from 0.05 to 0.76 cm year^{-1} (median of

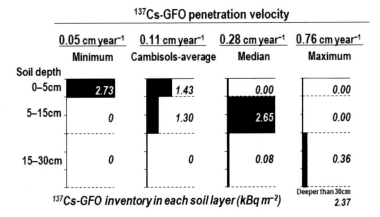

Fig. 20.4 Hypothetical vertical distribution of ^{137}Cs-GFO as of October 1, 2008, estimated by assuming constant vertical penetration velocities for the monthly deposition of ^{137}Cs in Tokyo. The penetration velocity was based on Jagercikova et al. (2015). The area of the black bar was proportional to the amount of inventory in each soil layer. Numbers in italics indicated the amount of the inventory. In assuming a penetration velocity of 0.76 cm year^{-1}, 2.37 kBq m^{-2} of ^{137}Cs was estimated to be present in the soil at more than 30 cm depth

0.28 cm year^{-1}) over 25 years, in which velocities for Cambisols, corresponding to a typical forest soil in Japan (brown forest soils), averaged 0.11 cm year^{-1} (Jagercikova et al. 2015). Applying these penetration velocities to the monthly deposits of ^{137}Cs in Tokyo, the vertical distribution of ^{137}Cs-GFO as of October 1, 2008, was estimated as shown in Fig. 20.4. For Cambisols-average (0.11 cm year^{-1}), the total inventory of 2.73 kBq m^{-2} was likely distributed in the 0–5 cm layer (1.43 kBq m^{-2}, 52% of total) and the 5–15 cm layer (1.30 kBq m^{-2}, 48%).

Here, we consider the intermediate velocity of the four vertical penetration velocities listed in the example (Fig. 20.4). Suppose we assume that ^{137}Cs-GFO mitigates only vertically through the soil at a constant infiltration rate. In that case, there are five possible vertical distributions: only in the first layer (0–5 cm), in the first and second (5–15 cm) layers, mainly in the second layer, in the second and third (15–30 cm) layers, and mainly in the third layer.

However, the ^{137}Cs-GFO vertical distribution patterns obtained from the actual data varied widely. We categorized 1136 soil pits with a complete three depth layer into a total of 15 patterns for the vertical distribution of ^{137}Cs-GFO based on the relative magnitude of its accumulation in the three soil layers (Fig. 20.5, see the footnote for details of the classification criteria). Among these, four patterns contained some percentage (more than 0.5% of the total inventory) of ^{137}Cs-GFO in all three layers (the fifth to eighth patterns from the left in Fig. 20.5). These four patterns accounted for 44% of the total soil pits. As mentioned in the previous paragraph, such patterns rarely appear under the assumption of vertical mitigation of ^{137}Cs-GFO at a constant infiltration rate. Additionally, we found several soil pits, 16% of the total, showing patterns where the amount of ^{137}Cs-GFO per unit depth was the smallest in the second layer (the four patterns noted on the right side of

Fig. 20.5). The infiltration-driven assumption, or at least the assumption of a constant infiltration rate, can hardly explain the existence of these patterns. In the next subsection, we examine whether the ^{137}Cs-GFO inventory in the soil pit showing each vertical distribution pattern is consistent with the expected magnitudes based on the infiltration-driven assumption.

20.5.2   The ^{137}Cs-GFO Inventory Predicted by Assuming Only Downward Infiltration in Soils Is Entirely Different from the Actual Value

We assume that the redistribution of ^{137}Cs-GFO was determined only by downward migration at a constant infiltration rate. The upper part of the schematic diagram shows relative magnitudes of infiltration rate and ^{137}Cs-GFO inventory expected from the vertical distribution pattern (Fig. 20.5).

In the four patterns on the far left (Fig. 20.5), the third layer and the second and third layers were "None." Based on the infiltration-driven assumption, even the earliest fallout of ^{137}Cs-GFO did not reach the second/third layer even after 63 years (i.e., even in 2008, 63 years after 1945) due to the very slow infiltration rate ("Very Late" to "Late" in expected infiltration, first line at the upper part of the schematic diagram, Fig. 20.5). Therefore, all the fallen ^{137}Cs-GFO would have remained in the soil within 15 or 5 cm of the surface layer, and the total inventory should be "Very Large" (expected inventory, second line at the upper part of the schematic diagram, Fig. 20.5). However, contrary to this expectation, the inventory for these four patterns was much smaller than that for the other patterns (0.62–2.01 kBq m^{-2}, the lower part of the schematic diagram, Fig. 20.5).

There were five patterns where the third layer was "All," "Most," or "More," indicating that the third layer had the largest concentration of ^{137}Cs-GFO (patterns of eighth, 10th, 11th, 14th, and 15th from the left, Fig. 20.5). In these five patterns, the infiltration rate must be so fast that a considerable amount of ^{137}Cs-GFO has already reached the third layer ("Very Fast" to "Fast" in expected infiltration, Fig. 20.5). It could be assumed that some of the ^{137}Cs-GFO has penetrated the soil below the third layer, deeper than 30 cm below the surface, and therefore the inventory of ^{137}Cs-GFO should be "Small" or "Very Small" (expected inventory, Fig. 20.5). However, partly contrary to this expectation, the inventory trends were split between two extremes: 2 of the 5 patterns had fairly small inventories, as expected, whereas 3 patterns were quite large, even the largest inventory of the 15 patterns (the eighth pattern from the left, Fig. 20.5).

We have concluded that the infiltration-driven assumption could not reasonably explain the ^{137}Cs-GFO-related combinations between the vertical distribution patterns and inventory quantities. In the next subsection, we will try to provide a consistent description of the ^{137}Cs-GFO-related combinations with respect to horizontal sediment migration on forested hillslopes.

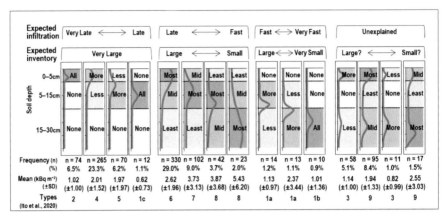

Fig. 20.5 Vertical distribution patterns of ^{137}Cs-GFO based on the relative magnitude of its accumulation in the three soil layers. Schematic diagram of the 15 patterns. To assess the relative accumulation of ^{137}Cs-GFO in each soil layer, we introduced two indices (Ito et al. 2020): the percentage of ^{137}Cs-GFO amount for each soil layer to the total of soil pit (P_n, %) for each nth soil layer, where n represents the order of the soil layer (1, 0–5 cm, 2, 5–15 cm; and 3, 15–30 cm depth), and the amount of ^{137}Cs-GFO per unit depth for each soil layer (Q_n, kBq m^{-2} cm^{-1}). We set the threshold to $P_n < 0.5\%$ for a layer having "None" ^{137}Cs-GFO accumulation. We extracted the layers for which the index P_n was determined to be "None." The other soil layer was determined to be "More" and "Less," or "Most," "Mid," and "Least," in order of increasing index Q_n. We categorized the vertical distribution pattern by combining these four statuses for the three layers of each soil pit. The bold gray curve is a hypothetical vertical distribution to intuitively understand the relative amounts of ^{137}Cs-GFO in the three layers. In the upper part of the schematic diagram, relative magnitudes of infiltration rate and inventory of ^{137}Cs-GFO are expected from the vertical distribution pattern, assuming that the vertical distribution of ^{137}Cs-GFO in soil depends only on the vertical infiltration of ^{137}Cs-GFO. The vertical distribution patterns were arranged based on the similarity of these expected relative magnitudes of expected infiltration and expected inventory. At the bottom of the schematic diagram, frequency (the number of occurrences and %) and mean value ($\pm$SD) of the ^{137}Cs-GFO (kBq m^{-2}) inventory were shown for each pattern. Mean values were simple arithmetic averages without normalization by GLM. The number of types (subtypes) presented in Ito et al. (2020) was shown in the lower-most column. In the original paper, the vertical distribution patterns were classified into nine types with three subtypes using the same two indices, but in this chapter, the patterns were subdivided into 15 patterns

20.5.3 The Sediment Redistribution Is Consistent with the Correspondence Between the Vertical Distribution Pattern and the Inventory

The GLM model for estimating ^{137}Cs-GFO constructed in Sect. 20.4 is mentioned again. The GLM model, which incorporated three fixed effects (total annual precipitation, region, and dataset category) and a random effect (NFSCI sampling sites), was significantly estimated ^{137}Cs-GFO ($p < 0.0001$). Applying the GLM analysis, the ^{137}Cs-GFO variability within the NFSCI sampling sites (58.8% of the total) was greater than that among the sites (41.2%). This considerable within-sites variation among ~4 soil pit repetitions within a 0.1 ha plot strongly indicated the large microscale heterogeneity of the ^{137}Cs-GFO distribution within a catchment (Ito et al. 2020). We hypothesized that this spatially uneven ^{137}Cs-GFO distribution was due to horizontal sediment migration within a forest catchment. Similarly, we hypothesized that the various vertical distribution patterns of ^{137}Cs-GFO mentioned in the previous subsection (Fig. 20.5) were also caused by horizontal sediment migration. Therefore, we introduced the vertical distribution pattern as an explanatory variable in the GLM described above and examined whether the accuracy of ^{137}Cs-GFO estimation could be improved.

We modified the GLM model by incorporating the vertical distribution pattern as an additional explanatory variable (hereafter, revised GLM model). The vertical distribution pattern had a significantly strong explanatory effect in the revised GLM model ($p < 0.0001$). In other words, under normalized conditions for the impact of the other explanatory variable, the ^{137}Cs-GFO inventory differed significantly by vertical distribution pattern. Moreover, the ^{137}Cs-GFO variability within the NFSCI sampling sites in the revised GLM model (25.8% of the total) was considerably less than that in the previous GLM model (58.8%). This indicates that the vertical distribution pattern effectively explained the within-forest variation in the ^{137}Cs-GFO inventory.

We calculated the LSM of the ^{137}Cs-GFO inventory for each vertical distribution pattern and examined the post hoc Tukey-HSD test using the revised GLM model. The 15 vertical distribution patterns were shown rearranged by the magnitude of the LSM value (Fig. 20.6). Generally, the pattern of high inventory was the accumulation of ^{137}Cs-GFO in all three layers (right side of Fig. 20.6). Alternatively, the pattern of low inventory was the lack of ^{137}Cs-GFO (i.e., "None") in any layer (left side of Fig. 20.6). The following Sects. (20.5.3.1–20.5.3.6) discuss sediment transportations that can be assumed from the combinations of vertical distribution patterns and sediment migration.

20.5.3.1 Pattern of Stable

The pattern with the greatest frequency of occurrence among the 15 vertical distribution patterns was the one with ^{137}Cs-GFO in all three layers and decreasing concentration from the surface layer to the bottom layer, accounting for 29% of the total (pattern 10, Fig. 20.6). We considered this pattern the most stable with respect to sediment migration, because it had the smallest difference in the amount of ^{137}Cs-GFO from the cumulative fallout in the revised GLM model (data not shown).

Although we named the pattern "stable," we cannot say for sure that there was no ^{137}Cs transfer at all. Tracer studies using ^{137}Cs have been accumulated as a method to clarify such long-term sediment transport (Ritchie and Ritchie 2005). This method is based on determining a reference site in the study area that can be assumed to have been stable without sediment disturbance, not only landslide and erosion but also rain splash transport and soil creep (Benda et al. 2005). This determination is performed with great care, because it determines the study's accuracy (Mabit et al. 2013; Fulajtar et al. 2017). The location of the soil pits used in this study was systematically determined, and it is unlikely that they would have happened to match any of the reference sites. The soil pits classified as "stable" in this study were not stable enough to be reference sites. It is appropriate to regard this "stable" pattern as a spot where significant erosion or accumulation did not appear in the vertical distribution patterns or as a "neutral" spot where erosion and accumulation were almost balanced.

20.5.3.2   Pattern of Erosion

Vertical distribution patterns 2 and 7 can be regarded as high erosion and erosion spots, respectively (Fig. 20.6). Pattern 7, at 23.3% of the total, was the second most common pattern. In these patterns, ^{137}Cs-GFO was distributed only in the surface layer or the surface and second layers, where the inventory was significantly smaller than that of pattern 10, considered "stable." As described in Sect. 20.5.2, the hypothesis that this pattern was due to a slow penetration rate is inconsistent with a very small inventory. Therefore, this situation can be explained by the fact that some surface-soil-adsorbing ^{137}Cs-GFO was eroded and transported elsewhere, and that the others remained in the spot. Forestry environments have been characterized by significant sediment redistribution in Japan (Gomi et al. 2008). This study also indicates that a large part of the forested area was subject to slow and mild erosion over a steep forested hillslope.

20.5.3.3   Pattern of Redeposition

The third most common pattern, pattern 13, 9.0% of the total, had ^{137}Cs-GFO in all three layers, with the second layer having the largest concentration, followed by the first, and the third layer being the smallest (pattern 13, Fig. 20.6). We consider this pattern to be the result of sediment redeposition. In other words, the sediment that eroded from the upslope area remained within the forest catchment instead of fluvial discharge.

It might be hypothesized that pattern 13 suggests being "stable" with faster infiltration rates than pattern 10, but this hypothesis could be rejected due to the large inventory of pattern 13. If the fast infiltration rate caused less ^{137}Cs-GFO in the first layer, it should also cause ^{137}Cs-GFO to infiltrate deeper than the third layer, resulting in a smaller inventory for the soil pits. However, the data showed the exact opposite. It makes more sense to consider that the accumulation of ^{137}Cs-GFO-rich soil (the most surface soils at the peak of fallout deposition) from the surrounding area led to a significantly large inventory for pattern 13.

Moreover, we consider patterns 14 and 15 to indicate high redeposition. These patterns represent extremely large inventories, although their frequency was low at 3.7% and 2.0% of the total. Comparing these three redeposition patterns (patterns 13–15), the inventory was larger for those with more ^{137}Cs-GFO in deeper layers. This indicates that the greater redeposited soil over 50 years, the greater the

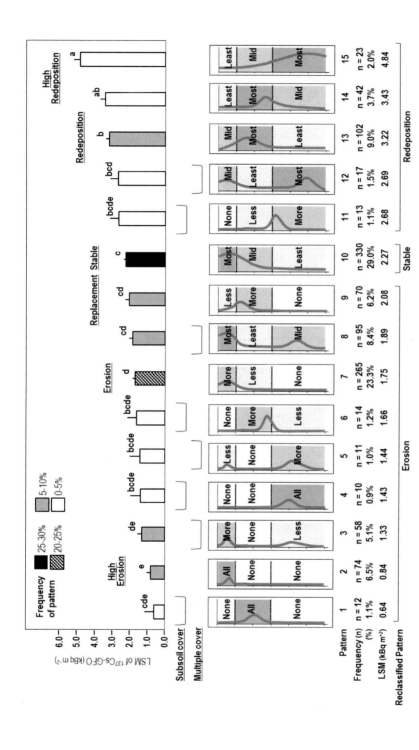

Fig. 20.6 Vertical distribution patterns of ^{137}Cs-GFO were rearranged by the ^{137}Cs-GFO inventory. Named pattern number, frequency (number of occurrences and %), the least-squares mean (LSM) of ^{137}Cs-GFO (kBq m^{-2}), and reclassified pattern (see Fig. 20.7) for each pattern were shown in the lower part of the schematic diagram. Bars and error bars show the LSM and standard error calculated by the GLM. The expected sediment migration (underlined text) was shown above or below the bar. The filled pattern in the bar indicates the frequency of occurrence. Post hoc comparisons between patterns (Tukey-HSD test) were indicated by letters above the bars showing the LSM of ^{137}Cs-GFO

accumulation of earlier eroded ^{137}Cs-GFO-rich soil in the deeper layers. Thus, it would be reasonable to expect a larger inventory in such a pattern.

The small concentration of ^{137}Cs-GFO in the top surface layer in patterns 14 and 15 can be reasonably explained by considering that the recently eroded soil from the surrounding area (i.e., soil redeposited in the top surface layer in patterns 14 and 15) did not contain much ^{137}Cs-GFO, because it was subsoil at the time of maximum fallout.

These "redeposition" patterns occupied 34 out of 66 (52%) soil pits with relatively high inventory (>6 kBq m^{-2}), which corresponded to a long-tail of a right-skewed frequency distribution of inventory on a soil pit basis (see Sect. 20.3, Fig. 20.2c). In particular, soil pits with outstandingly high inventory (top six soil pits with >14 kBq m^{-2}, Fig. 20.2c) belonged to patterns 13–15 without exception.

In these "(High) Redeposition" patterns, the accumulation of ^{137}Cs-GFO may not be limited to the surface layer of 30 cm but could be deeper. This situation can occur if the sediment redeposition over 50 years is more than 30 cm thick.

20.5.3.4 Pattern of Replacement

Pattern 9 had "Less" in the first layer, "More" in the second layer, and "None" in the third layer (Fig. 20.6). The frequency was relatively high, accounting for 6.2% of the total. The inventory (2.08 kBq m^{-2}) was between "Stable" (pattern 10) and "Erosion" (pattern 7) and not significantly different from either of them. We considered this pattern as indicating sediment replacement along the slope transect. A relatively slow and prolonged sediment redistribution with downward transport of ^{137}Cs-GFO-rich soil and redepositing ^{137}Cs-GFO-poor soil upward could result in the less ^{137}Cs-GFO in the top layer and the inventory being slightly smaller than the "stable" pattern. Repeated sheet wash erosion and soil creeping over the years may have contributed to this soil replacement.

20.5.3.5 Patterns of Subsoil Cover

Patterns 1, 4, 6, and 11 had the presence of the "None" layer in common, that is, soil layer containing very little ^{137}Cs-GFO, on the surface (Fig. 20.6). The occurrence of these patterns was not very high, with a range of 0.9–1.2%. The total rate of occurrence of the four patterns was 4.3%. Due to the small number of occurrences, there was no significant difference in the inventory between these four patterns and any of patterns 2 (High Erosion), 7 (Erosion), 10 (Stable), and 13 (Redeposition), except for between patterns 1 and 13.

These patterns were considered spots where at least 5 cm of soil containing little ^{137}Cs-GFO from the surrounding area was redeposited. It was assumed that the redeposited soil was a subsoil at the time of ^{137}Cs-GFO deposition. It was impossible to determine whether this redeposition of subsoil occurred over a long period or was caused by a single event such as a landslide (or possibly anthropogenic events such as road construction). However, it can be considered a considerable mass movement.

The inventory of these patterns ranged from very small (0.64 kBq m^{-2}, pattern 1) to larger than "Stable" (2.68 kBq m^{-2}, pattern 11). This can be inferred as a difference in the conditions before the subsoil redeposition. Erosion may have occurred even before subsoil accumulation in pattern 1, and continuous redeposition of topsoils may have occurred in pattern 11.

20.5.3.6 Patterns of Multiple Covers

The remaining four patterns (3, 5, 8, and 12) had the second layer as "None" or "Least" in common (Fig. 20.6). ^{137}Cs-poor soil was likely redeposited first, and then, ^{137}Cs-rich soil was redeposited (but there may be other processes). This can be rephrased as patterns that experienced multiple sediment transport events with different sources of soils.

The inventory of these patterns was as widely ranged as the patterns of subsoil cover (Sect. 20.5.3.5). The occurrence rates of these patterns (3, 5, 8, and 12) were 5.1%, 1.0%, 8.4%, and 1.5%, respectively. The total occurrence of the four patterns was 16.0% of the total, much higher than the total of patterns of subsoil cover (4.3%).

20.5.4 Whole Picture of ^{137}Cs-GFO-Containing Sediment Migration

We classified the vertical distribution patterns in the previous Sect. 20.5.3 into six categories with respect to sediment transport: stable (pattern 10, 29.0%), erosion (pattern 7, 23.3%), (high) redeposition (patterns 13–15, 14.7%), replacement (pattern 9, 6.2%), subsoil cover (patterns 1, 4, 6, 11, 4.3%), and multiple cover (patterns 3, 5, 8, and 12, 16.0%). In this section, we will aggregate the vertical distribution pattern with respect to gain and loss of ^{137}Cs-GFO.

The stable pattern (pattern 10, 29.0%) was used as the standard. Patterns 1–9 with less ^{137}Cs-GFO inventory were reclassified as patterns of erosion, and patterns 11–15 with more ^{137}Cs-GFO inventory were reclassified as patterns of redeposition. We further revised the GLM model, incorporating the reclassified pattern (i.e., stable, erosion, and redeposition) as an explanatory variable instead of the 15 vertical distribution patterns (patterns 1–15, Fig. 20.6). This further revised GLM model (i.e., a GLM model with reclassified pattern, total annual precipitation, region, and dataset category as fixed effects and NFSCI sampling sites as a random effect) determined the LSM values for each reclassified pattern (Fig. 20.7). These LSM values were ^{137}Cs-GFO inventory estimates in a hypothetical location normalized for other fixed and random effects associated with the total amount of the ^{137}Cs-GFO

Fig. 20.7 Vertical distribution patterns of ^{137}Cs-GFO aggregated by the redistribution of ^{137}Cs-GFO. Bars and error bars show the LSM and 95% confidence intervals, respectively. The pie chart shows the frequency of the pattern

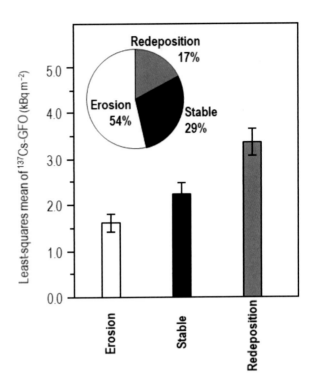

deposition. The LSMs suggest that the difference between stable and redeposition was slightly less than twice the difference between stable and erosion (Fig. 20.7).

The frequency of occurrence of the three patterns is also shown in Fig. 20.7. It is suggested that more than 50% of the forest area was reduced in ^{137}Cs-GFO due to sediment erosion, and that the area with enhanced ^{137}Cs-GFO due to sediment accumulation was approximately 1/3 of the area of erosion. Thus, more than 70% of the total forested area (i.e., the sum of the erosion and redeposition patterns) was considered altered by these sediment migrations. This suggests the frequency of sediment transport in forests on the Japanese archipelago over the past 50 years. To summarize, a general pattern of sediment migration within a forest catchment became apparent, that is, thinly eroded sediment from a large area accumulates thickly in a narrow area.

20.6 Destination of ^{137}Cs-GFO in Forested Areas Across Japan After Half a Century from Fallout Deposition

This study empirically demonstrated that most ^{137}Cs-GFO in the forests remained in the surface soil after several decades of fallout. Precisely, we found no significant difference between the inventory and cumulative fallout in the GLM model (Sect.

20.4, Ito et al. 2020). However, the GLM analysis also showed that the estimated inventory was only 79% of the cumulative fallout. Where does the remaining 21% of the ^{137}Cs-GFO reside other than in the 0–30 cm of the forest soil surface?

The first possibility is sediment discharges into stream water, that is, ^{137}Cs-GFO loss via fluvial discharge from forest catchments. The reported annual ^{137}Cs runoff of the total deposition of the forest catchment after the FDNPP accident was less than 0.3% (Onda et al. 2020). The yearly runoff rate of 0.3% year^{-1} was equivalent to 12.8% of the accumulated ^{137}Cs-GFO discharge percentage over 50 years (provisionally calculated using MRI observation data, see Fig. 20.1). This may be an overestimate or an underestimate. First, the measurements based on the runoff rate immediately after the FDNPP accident are potentially overestimated, because the annual ^{137}Cs runoff rate decreases exponentially over time (Kato et al. 2017). However, as mentioned in the next section, there were many sediment-related disasters during the peak of the GFO fallout, so the runoff rate at that time might have been higher than the 0.3% year^{-1} used in the estimation.

The second possibility is that some ^{137}Cs-GFO was pooled locally in the deep soil. As shown in Sect. 20.5.3.3, there were soil pits with a very large inventory, although the number of occurrences was small. In those high redeposition patterns (patterns 14–15, Fig. 20.6), more ^{137}Cs-GFO was found in the deeper layers, suggesting a high possibility that ^{137}Cs-GFO was also pooled in deeper layers than 30 cm.

The percentage of these two possible ^{137}Cs-GFO destinations is unknown. Clarifying the ^{137}Cs-GFO inventory in deeper soils at the high redeposition sites will provide more reliable estimates.

20.7   Sediment Migration and Past Forest Conditions: Suggestions for the Future

This study clearly showed that sediment migration within the forest catchment accompanied by spatial redistribution of ^{137}Cs-GFO had frequently been occurring in the past half-century. Consequently, the ^{137}Cs-GFO inventory becomes spatially heterogeneous in the forest. In the early 1960s, when global fallout peaked, Japanese forests were recovering from extensive deforestation during and after World War II. Overlogging of forests induces soil erosion (García-Ruiz et al. 2017), and the effects continue for ~20 years after the logging pressure decreases (Tada 2018). In Japan, sediment-related disasters frequently occurred from 1950 to 1985 (Tada 2018). During the same period, minor sediment migration that was not counted as landslides must have occurred frequently. The periods of the global fallout coincided with the frequent occurrence of sediment migration. In essence, the forest management history over the past half-century may be closely related to the ^{137}Cs-GFO redistribution.

This study demonstrated that most ^{137}Cs-GFO remained in the forest even after experiencing a period of poor-forested conditions. If forests can be appropriately managed in the future (in other words, without falling below past levels), the discharge of FDNPP-generated ^{137}Cs into aquatic ecosystems could be suppressed.

Topsoil discharge is considerably influenced by the forest floor cover condition caused by the forest understory vegetation and litter layer. Forest management is necessary to prevent the forest floor from becoming bare of the mineral soil layer to reduce the discharge. Specifically, it is essential to conduct appropriate thinning in planted forest areas to create a light environment on the forest floor that allows forest floor vegetation to cover the ground surface (Fig. 20.8). Additionally, as the density of Japanese deer increases, the decline of forest floor vegetation has been a severe problem throughout Japan, and it is essential to address this issue (Fig. 20.9). Even clear-cutting, one of the most severe disturbances in forest management, has been shown to have only a temporary effect in promoting the discharge of FDNPP-generated [137]Cs (Nishikiori et al. 2019). Implementing long-term proactive forest

Fig. 20.8 Forest floors of hinoki cypress (*Chamaecyparis obtusa*) forests. (**a**) A forest floor with exposed mineral soils under limited light environment without timely thinning. (**b**) Extreme soil erosion. (**c**) Soil pillars created with rain splash erosion on the forest floor (the lower-left bar is approximately 10 cm). (**d**) Vegetation-covered forest floor under improved light conditions due to thinning

Fig. 20.9 Forest floor of forest areas with a high density of sika deer
(**a**) A birch forest floor dominated by unpalatable ferns. (**b**) A clear-cut evergreen oak forest floor with bare mineral soils and a few unpalatable shrubs. The evergreen oak forest details are shown elsewhere (Suzuki and Ito 2014)

management in the affected areas is necessary to prevent sediment discharge from the forest watershed.

Finally, the study results were achieved using archived soil samples obtained from nationwide systematic sampling. The fallout observation data that the Japan Meteorological Agency and local authorities have maintained for many years also provided essential information for this study. There are many examples, not limited to this study, where the existence of scientifically reliable samples and specimens contributed to the resolution of issues that were not envisioned when they were collected. To meet future demands, we believe that it is essential to continue observations and sample management in various fields of science.

Acknowledgements We thank the Department of Forest Soils of Forestry and Forest Products Research Institute for providing soil samples and support for sample preparations and analysis. We are grateful to Prof. T. M. Nakanishi and Prof. K. Tanoi for their support for the initial improvement of radioactivity analysis using an NaI (Tl) scintillation counter. We express our deep sympathy for the loss of our highly talented computing assistant Dr. Masato Ito, FFPRI. This work was partly financially assisted by KAKENHI (No. 25292099) from the Ministry of Education, Culture, Sports, Science and Technology (MEXT) of Japan.

References

Aoyama M (1999) Geochemical studies on behavior of anthropogenic radionuclides in the atmosphere. PhD thesis. Kanazawa University, Kanazawa
Aoyama M (2018) Long-range transport of radiocaesium derived from global fallout and the Fukushima accident in the Pacific Ocean since 1953 through 2017—part I: source term and

surface transport. J Radioanal Nucl Chem 318:1519–1542. https://doi.org/10.1007/s10967-018-6244-z

Aoyama M (2019) The integrated global fallout database—IGFD. University of Tsukuba, Tsukuba. https://doi.org/10.34355/CRiED.U.Tsukuba.00005

Aoyama M, Hirose K, Suzuki Y, Inoue H, Sugimura Y (1986) High level radioactive nuclides in Japan in May. Nature 321:819–820. https://doi.org/10.1038/321819a0

Aoyama M, Hirose K, Sugimura Y (1987) Deposition of gamma-emitting nuclides in Japan after the reactor-IV accident at Chernobyl. J Radioanal Nucl Chem 116:291–306. https://doi.org/10.1007/BF02035773

Benda L, Hassan MA, Church M, May CL (2005) Geomorphology of steepland headwaters: the transition from hillslopes to channels. J Am Water Resour Assoc 41:835–851. https://doi.org/10.1111/j.1752-1688.2005.tb03773.x

Dyer A, Chow JK, Umar IM (2000) The uptake of caesium and strontium radioisotopes onto clays. J Mater Chem 10:2734–2740. https://doi.org/10.1039/B006662L

Fukuyama T, Takenaka C, Onda Y (2005) ^{137}Cs loss via soil erosion from a mountainous headwater catchment in central Japan. Sci Total Environ 350:238–247. https://doi.org/10.1016/j.scitotenv.2005.01.046

Fulajtar E, Mabit L, Renschler CS, Lee ZY (2017) Use of ^{137}Cs for soil erosion assessment. Food and Agriculture Organization of the United Nations (FAO), Rome. https://www.fao.org/3/i8211e/i8211e.pdf

García-Ruiz JM, Beguería S, Arnáez J, Sanjuán Y, Lana-Renault N, Gómez-Villar A, Álvarez-Martínez J, Coba-Pérez P (2017) Deforestation induces shallow landsliding in the montane and subalpine belts of the Urbión Mountains, Iberian range, Northern Spain. Geomorphology 296:31–44. https://doi.org/10.1016/j.geomorph.2017.08.016

Gomi T, Sidle RC, Miyata S, Kosugi K, Onda Y (2008) Dynamic runoff connectivity of overland flow on steep forested hillslopes: scale effects and runoff transfer. Water Resour Res 44:W08411. https://doi.org/10.1029/2007WR005894

Higuchi H, Fukatsu H, Hashimoto T, Nonaka N, Yoshimizu K, Omine M, Takano N, Abe T (1988) Radioactivity in surface air and precipitation in Japan after the Chernobyl accident. J Environ Radioact 6:131–144. https://doi.org/10.1016/0265-931X(88)90056-2

Hirose K, Aoyama M, Katsuragi Y, Sugimura Y (1987) Annual deposition of Sr-90, Cs-137 and Pu-239, 240 from the 1961–1980 nuclear explosions: a simple model. J Meteorol Soc Jpn 65:259–277. https://doi.org/10.2151/jmsj1965.65.2_259

Ito E, Miura S, Aoyama M, Shichi K (2020) Global ^{137}Cs fallout inventories of forest soil across Japan and their consequences half a century later. J Environ Radioact 225:106421. https://doi.org/10.1016/j.jenvrad.2020.106421

Iwagami S, Onda Y, Tsujimura M, Abe Y (2017) Contribution of radioactive ^{137}Cs discharge by suspended sediment, coarse organic matter, and dissolved fraction from a headwater catchment in Fukushima after the Fukushima Dai-ichi Nuclear Power Plant accident. J Environ Radioact 166:466–474. https://doi.org/10.1016/j.jenvrad.2016.07.025

Jagercikova M, Cornu S, Le Bas C, Evrard O (2015) Vertical distributions of ^{137}Cs in soils: a meta-analysis. J Soils Sediments 15:81–95. https://doi.org/10.1007/s11368-014-0982-5

Kato H, Onda Y, Hisadome K, Loffredo N, Kawamori A (2017) Temporal changes in radiocesium deposition in various forest stands following the Fukushima Dai-ichi Nuclear Power Plant accident. J Environ Radioact 166:449–457. https://doi.org/10.1016/j.jenvrad.2015.04.016

Katsuragi Y (1983) A study of ^{90}Sr fallout in Japan. Pap Meteorol Geophys 33:277–291. https://doi.org/10.2467/mripapers.33.277

Mabit L, Meusburger K, Fulajtar E, Alewell C (2013) The usefulness of ^{137}Cs as a tracer for soil erosion assessment: a critical reply to Parsons and Foster (2011). Earth Sci Rev 127:300–307. https://doi.org/10.1016/j.earscirev.2013.05.008

Malakhov S, Pudovkina I (1970) Strontium 90 fallout distribution at middle latitudes of the northern and southern hemispheres and its relation to precipitation. J Geophys Res 75:3623–3628. https://doi.org/10.1029/JC075i018p03623

Nanko K, Hashimoto S, Miura S, Ishizuka S, Sakai Y, Levia DF, Ugawa S, Nishizono T, Kitahara F, Osone Y, Kaneko S (2017) Assessment of soil group, site and climatic effects on soil organic carbon stocks of topsoil in Japanese forests. Eur J Soil Sci 68:547–558. https://doi.org/10.1111/ejss.12444

Nishikiori T, Hayashi S, Watanabe M, Yasutaka T (2019) Impact of clearcutting on radiocesium export from a Japanese forested catchment following the Fukushima nuclear accident. PLoS One 14:e0212348. https://doi.org/10.1371/journal.pone.0212348

Omori Y, Hosoda M, Takahashi F, Sanada T, Hirao S, Ono K, Furukawa M (2020) Japanese population dose from natural radiation. J Radiol Prot 40:R99–R140. https://doi.org/10.1088/1361-6498/ab73b1

Onda Y, Taniguchi K, Yoshimura K, Kato H, Takahashi J, Wakiyama Y, Coppin F, Smith H (2020) Radionuclides from the Fukushima Daiichi Nuclear Power Plant in terrestrial systems. Nat Rev Earth Environ 1:644–660. https://doi.org/10.1038/s43017-020-0099-x

Rafferty B, Brennan M, Dawson D, Dowding D (2000) Mechanisms of ^{137}Cs migration in coniferous forest soils. J Environ Radioact 48:131–143. https://doi.org/10.1016/S0265-931X(99)00027-2

Ritchie JC, Ritchie CA (2005) Bibliography of publications of ^{137}cesium studies related to erosion and sediment deposition. USDA-Agricultural Research Service, Beltsville. https://hrsl.ba.ars.usda.gov/cesium/Cesium137bib.htm

Satoh D, Furuta T, Takahashi F, Endo A, Lee C, Bolch WE (2014) Calculation of dose conversion coefficients for external exposure to radioactive cesium distributed in soil. JAEA Res 2014:017. https://doi.org/10.11484/JAEA-RESEARCH-2014-017

Shinomiya Y, Tamai K, Kobayashi M, Ohnuki Y, Shimizu T, Iida S, Nobuhiro T, Sawano S, Tsuboyama Y, Hiruta T (2014) Radioactive cesium discharge in stream water from a small watershed in forested headwaters during a typhoon flood event. Soil Sci Plant Nutr 60:765–771. https://doi.org/10.1080/00380768.2014.949852

Suzuki M, Ito E (2014) Combined effects of gap creation and deer exclusion on restoration of belowground systems of secondary woodlands: a field experiment in warm-temperate monsoon Asia. For Ecol Manag 329:227–236. https://doi.org/10.1016/j.foreco.2014.06.028

Tada Y (2018) Landscape changes and disasters in Japan. Water Sci 62:121–137. https://doi.org/10.20820/suirikagaku.62.4_121

Taniguchi K, Onda Y, Smith HG, Blake W, Yoshimura K, Yamashiki Y, Kuramoto T, Saito K (2019) Transport and redistribution of radiocesium in Fukushima fallout through rivers. Environ Sci Technol 53:12339–12347. https://doi.org/10.1021/acs.est.9b02890

Tsuji H, Nishikiori T, Yasutaka T, Watanabe M, Ito S, Hayashi S (2016) Behavior of dissolved radiocesium in river water in a forested watershed in Fukushima prefecture. J Geophys Res Biogeosci 121:2588–2599. https://doi.org/10.1002/2016JG003428

Ueda S, Hasegawa H, Kakiuchi H, Akata N, Ohtsuka Y, Hisamatsu S (2013) Fluvial discharges of radiocaesium from watersheds contaminated by the Fukushima Dai-ichi nuclear power plant accident, Japan. J Environ Radioact 118:96–104. https://doi.org/10.1016/j.jenvrad.2012.11.009

Ugawa S, Takahashi M, Morisada K, Takeuchi M, Matsuura Y, Yoshinaga S, Araki M, Tanaka N, Ikeda S, Miura S (2012) Carbon stocks of dead wood, litter, and soil in the forest sector of Japan: general description of the National Forest Soil Carbon Inventory. Bull For For Prod Res Inst 425:207–221. https://www.ffpri.affrc.go.jp/pubs/bulletin/425/documents/425-2.pdf

UNSCEAR (2020) UNSCEAR 2020/2021 Report: "Sources, effects and risks of ionizing radiation". https://www.unscear.org/unscear/en/publications.html

Yoshimatsu H, Abe S (2006) A review of landslide hazards in Japan and assessment of their susceptibility using an analytical hierarchic process (AHP) method. Landslides 3:149–158. https://doi.org/10.1007/s10346-005-0031-y

Zhiyanski M, Bech J, Sokolovska M, Lucot E, Bech J, Badot PM (2008) Cs-137 distribution in forest floor and surface soil layers from two mountainous regions in Bulgaria. J Geochem Explor 96:256–266. https://doi.org/10.1016/j.gexplo.2007.04.010

Chapter 21
Resilience Education Program in Iitate Village for the Young Generation

Hiroaki Sugino and Masaru Mizoguchi

21.1 Introduction

The Great East Japan Earthquake of 2011 resulted in not only the destruction of cities and villages but also the spread of radioactive substances due to the accident in the Fukushima Daiichi Nuclear Power Plant. Iitate village was one of the municipalities affected by the accident despite being located 30–45 km from the power plant, since the Northwestern winds were blowing at the time of the accident. Although the order to evacuate Iitate was later compared to other evacuation orders, it was issued much later than others, on May 15, 2011. As a result of the enormous efforts of the local municipalities, residents, and volunteers to decontaminate and restore the affected areas (e.g., Mizoguchi 2012, 2013; Resurrection of Fukushima 2013), almost all evacuation orders were lifted from the municipalities in Fukushima, including Iitate by 2022 (Table 21.1).

Currently, residential functions have been restored in almost all districts of Iitate except for Nagadoro district, which is still designated as a "difficult-to-return zone," which refers to areas where the annual integral air dose estimation has not reduced to less than 20 mSv (Fig. 21.1). Originally, Iitate was rich in natural resources (75% of the village is mountainous forest) and was selected as one of the Most Beautiful Villages of Japan in 2010. At present, despite the relatively high elevation and snowfall and other challenges to cultivation, agricultural revival in the area has focused on attempting to cultivate rice, colorful flowers, and vegetables that are well suited to the difference in temperature between day and night (Fig. 21.2).

Eleven years after the disaster, although physical reconstruction, including infrastructure development, is on track, approximately 50% of residents still feel that Fukushima is yet to be sufficiently rebuilt (Government survey conducted in 2021).

H. Sugino (✉) · M. Mizoguchi
Graduate School of Agricultural and Life Sciences, The University of Tokyo, Tokyo, Japan
e-mail: a-sugino@g.ecc.u-tokyo.ac.jp

T. M. Nakanishi, K. Tanoi (eds.), *Agricultural Implications of Fukushima Nuclear Accident (IV)*, https://doi.org/10.1007/978-981-19-9361-9_21

Table 21.1 Municipalities in which decontamination was completed and evacuation orders were lifted

Municipality	Evacuation order was lifted on	Completion of decontamination
Tamura city	April 1, 2014	June 2013
Kawauchi village	October 1, 2014[a]/June 14, 2016[b]	March 2014
Naraha town	September 5, 2015	March 2014
Katsurao village	June 12, 2016	December 2015
Minamisoma city	July 12, 2016	March 2017
Iitate village	March 31, 2017	December 2016
Kawamata village	March 31, 2017	December 2015
Namie town	March 31, 2017	March 2017
Tomioka town	April 1, 2017	January 2017
Okuma town	April 10, 2019	March 2014
Futaba town	March 4, 2020	March 2016

Reference: Website about decontamination published by the Ministry of the Environment (http://josen.env.go.jp/en/decontamination/)
[a]Former evacuation order lift prepared area
[b]Former residency restricted area

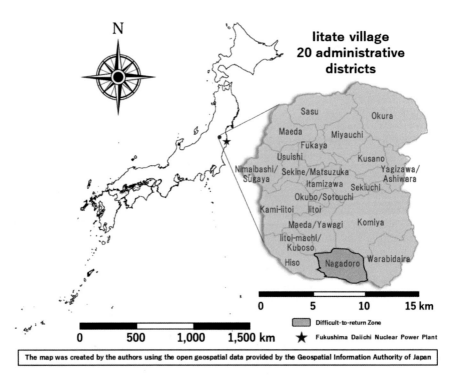

Fig. 21.1 Location of Iitate village and 20 administrative districts

Fig. 21.2   Agricultural products (rice, baby's breath, and napa cabbage) in Iitate (Source: authors)

Moreover, with village residents gradually returning and population and activity increasing, new challenges arise daily that need to be resolved. Through our ongoing field activities that have continued from 3 months after the accident in 2011 to the present (Mizoguchi 2016), we have observed that excursions and field trips to the affected areas qualitatively changed the awareness of participants, including students (e.g., Osada 2013; Nishiwaki et al. 2018; Ebitani et al. 2021). Further, we have come to believe that the resilience of the village, which has the disadvantage of being on land contaminated by radiation, must go beyond mere technical decontamination and physical reconstruction of infrastructure, and that it is important to educate young people to take up the challenge of creating a new style agriculture.

To pass on the knowledge of what led to the disaster and the lessons learned, it is becoming increasingly important to educate the next generation, who can discover and create solutions for the problems and challenges that constantly arise in agriculture. To this end, a specific learning program should be designed. The following sections will introduce the education program developed in Iitate village, Fukushima, and discuss the results of the qualitative and quantitative research on the feedback from the program participants.

21.2   FPBL: Field- and Project-Based Learning

It is difficult to restore the areas, communities, and agricultural land affected by the nuclear accident by only using specific science and technology in a segmented way. To achieve both physical reconstruction—such as infrastructure and decontamination—and social reconstruction, it is essential to create opportunities for the generation that has been actively involved in reconstruction to interact with younger generations inside and outside the affected areas and collaborate with them to solve issues in real time. It is also important to develop human resources who can discover what they can contribute to reconstruction and development. This goal is not limited to the field of agriculture but is also a challenge for the field of pedagogy. In the following sections, we will briefly describe the educational philosophy that served as the foundation for field- and project-based learning (FPBL), which is introduced and described in this chapter.

21.2.1 PBL: Project-Based Learning

Project-based learning (PBL) is the core concept of FPBL. PBL is a learning theory proposed by the American educationalist John Dewey and refers to a program and process of "learning to solve complex questions and hypotheses about the real world as a project program and its process" (Bender 2012). PBL has also been referred to as "Problem-based learning" (Wood 2003). Under PBL, students or participants learn through active effort to solve issues and problems by setting up hypotheses or research questions from a practical level. Hands-on activities and discussions help individual learning to more efficiently produce creative responses (Redish 2012). Such methods complement traditional input-based learning, which emphasizes the acquisition of information and skills through classroom lectures. By encouraging learners to take an active role in solving the problems set by the project, PBL can be expected to change attitudes toward the future, increase motivation (Sakai et al. 2020), and help participating students to gain a multifaceted understanding of the subject and become more proactive in their learning (Akui et al. 2019).

21.2.2 Open Spiral Model

The Open Spiral Model, a concept proposed by Hiroshi Harashima, advocates opening research and its results and technologies to society to connect academic and social communities. He warned that the gap among specialized knowledge, general knowledge, and terminology can make it difficult to disseminate the technology and value of research to society (Harashima 2016). To overcome this difficulty, he advocated a new relationship between research and society, where research is opened to society on a case-by-case basis and feedback is obtained directly and instantly from society or industry, instead of the traditional relationship where research is linearly connected to society through academia and industry (Fig. 21.3). The FPBL proposed in this chapter also adopts this Open Spiral Model, in which researchers are presented with practical issues by residents and farmers in the affected communities and farmlands, implement experiments and practices in the field, and receive immediate feedback to evolve research and projects.

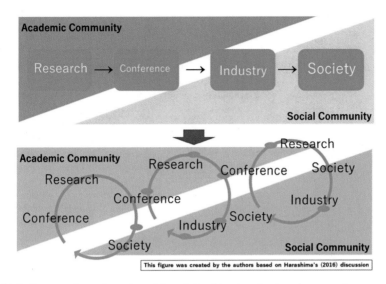

Fig. 21.3 Conceptual transformation of the relationship of academic and community

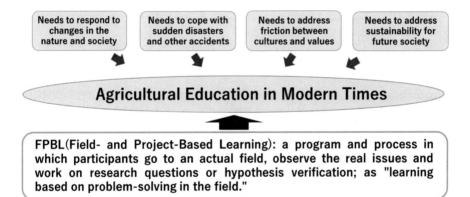

Fig. 21.4 Needs of agricultural education in modern times

21.2.3 *Multidisciplinary Science and STE(A)M*

Agricultural education today is a complex science, and solving problems in the field with a single specialty often proves extremely difficult. This is because the complex needs of the field continue to arise under changing conditions: (1) to respond to changes in the natural environment and society, (2) to cope with sudden natural disasters and other burdens, (3) to deal with frictions in culture and values, and (4) to ensure sustainability for the future (Fig. 21.4). In response to this, agricultural education needs to clarify the significance of continued focus on the agricultural field, to convey the importance of continuing to discover issues on one's own, and to

Table 21.2 Perspectives on FPBL and comparison with SBL and PBL

Comparison items	Subject-based learning (SBL)	Project-based learning (PBL)	**Field and project-based learning (FPBL)**
Format	Input model	Problem-solving model	**Problem-solving model in a real field**
Learning process	Basics → application	Hypothesis making and verification	**Repetition of hypothesis making and verification in a real field**
Answer	Only 1 prepared	Several answers	**Configure and adjust to the field while exploring multiple possibilities**
Purpose	Lead to the prepared answer	Lead to the solution for the problem	**Entire process from problem finding to implementation of the solution**
Learner	1 Person	1 Person ~ group	**Group with members of various academic fields and stakeholders**
Method	Blackboards and textbooks	Discussion	**Trial and error in the field**
Fieldwork	None	In some cases	**Mandatory**

foster a positive attitude toward the repetitive process of hypothesis generation, hypothesis testing, trial and error, and adjustment, in addition to the transfer of specialized knowledge. Fostering such practical and complex comprehensive abilities has been the aim of MST (Mathematics, Science, Technology) education in the 1990s and STEM (Science, Technology, Engineering and Mathematics) (STEM Education Act of 2015 (H.R. 1020, 114-56) (2015)) or STEAM (Science, Technology, Engineering, "Arts," and Mathematics) (e.g., Yakman 2008; Maeda 2013) education proposed in the late 2000s in the USA. They were originally proposed as a way to develop human resources with holistic sensibilities and innovative ideas and attitudes (Allina 2018), but this particular perspective is also needed for agricultural education in the modern era.

Integrating the concepts and discussions introduced above, we have proposed and are practicing the FPBL, a learning program that combines PBL with a "field-based" approach (Table 21.2). FPBL simply refers to a program and process in which participants go to the real field, observe the actual issues, and work on research questions or hypothesis verification, as learning based on problem-solving in the field. As a form of learning, it is different from conventional input-based learning, which emphasizes the acquisition of information and skills through classroom learning, and in that FPBL starts with the identification of issues from practical situations and aims to apply the results research to the field. In addition, it aims to develop human resources who can extract issues from the field, think, discuss, and work in a group of diverse and multidisciplinary people, and finally return meaningful results to the field.

21.3 Program and Activities in FPBL Practices

With Iitate as a practical field for FPBL, throughout the year, we carry out a cluster of activities that fall into 4 main categories: laboratory, course-work, collaborative, and student led (Fig. 21.5). We implement a hybrid program of education and research to drive the activities as shown in Fig. 21.6.

As a research program, we aim to foster international-level transdisciplinary research with resilience knowledge and ICT technologies, and to build a foundation for the transmission of a new Japanese-style (small-scale intergenerational) agriculture in Iitate. Specifically, we have conducted tailor-made composting experiments adapted to local field conditions, experiments on radio-wave transmission in mountainous forests, and research on measures against animal damage in farming.

In addition, as an educational program, we have conducted field trips and workshops in Iitate for university students to interact with agricultural practitioners and create opportunities to integrate the academic knowledge and local reconstruction/resilience knowledge for problem-solving and research activities in the field. We have also established and operated field museums, held online information sharing sessions (Table 21.3), and conducted surveys to assess the program's effectiveness.

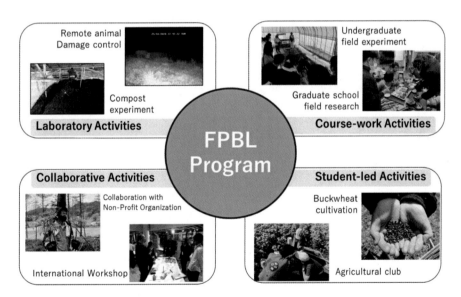

Fig. 21.5 4 Categories of FPBL program

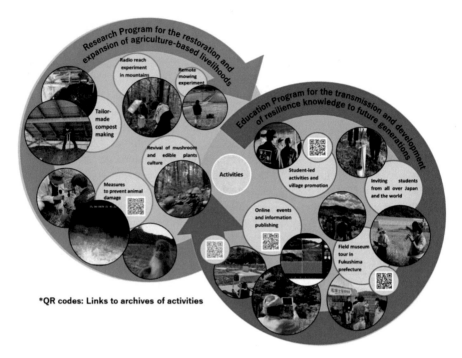

Fig. 21.6 Structure of hybrid program of education and research

Table 21.3 Major examples of online information transmission/publication

• Collection of Mizo's works on Fukushima (in English) *Mizo = one of the authors
http://www.iai.ga.a.u-tokyo.ac.jp/mizo/edrp/fukushima/Fukushima_articles.html
• Madei University project website
http://madeiuniv.jp/jinzai/
• Resilience Agronomy Starting from Fukushima
https://www.youtube.com/watch?v=XGT2i7eMgYQ
• Resilience Education Program in Iitate Village for the Young Generation
https://www.youtube.com/watch?v=e9tpZuEu3HE
• Archive of online flower watching tour 2022
https://www.youtube.com/watch?v=W8YaeqEfKww&t=267s
• Archive of online rice planting event 2021
https://www.youtube.com/watch?v=5KU_2p-eGsc&t=9688s
• Archive of International Online Joint Workshop held on December 7, 2021
https://www.youtube.com/watch?v=NJ00gO0kQhs
• Made in Iitate (Iitate village promotion video created by graduate students)
https://www.youtube.com/watch?v=l4Eu9_CeSz4
• Komiya Short Walk
https://www.youtube.com/watch?v=JKJokpyF_0k&t=1s

21.4 Effectiveness of FPBL

This section introduces a summary of the results of the feedback survey conducted before and after study tours. The targeted feedback surveys were conducted in conjunction with study tours conducted in Iitate village on a total of 16 occasions between 2019 and 2021. A total of 97 responses from both the pre- and postsurveys, excluding those that were incomplete, were considered valid responses. The structure of the questions obtained and analyzed is shown in Table 21.4.

The ultimate goal of the FPBL program is the enhancement of the ability to identify cutting-edge academic challenges in the field and the formation and development of sustainable and resilient communities. To this end, the level of awareness and activeness could be evaluated as a measure of the effectiveness of the program. Here, we focus on the concepts of "regional attachment" [an extension of Low and Altman's concept of "place attachment" (Low and Altman 1992)] and "self-efficacy," which have recently attracted attention as factors contributing to active participation in activities that make community contributions.

"Regional attachment" is defined as an emotional connection between people and places (Hidalgo and Hernandez 2001). Previous studies have shown that this concept plays an important role in cooperation in activities such as community development, environmental conservation, and disaster prevention (e.g., Brown et al. 2003; Ishimori 2004; Suzuki and Fujii 2008). Further, among the factors that determine human behavior (antecedents, consequences, and cognitive factors), "self-efficacy" (Bandura 1977) focuses on cognitive factors (cognition-based motivation and cognitive representations of entailment). It refers to the expectation of "how well one can perform the necessary behavior to produce a certain result" (Sakano 1986) and has been used in varied fields, such as education, industry, and preventive medicine, since it is a cognitive variable that both precedes and causes behavioral change (e.g., Ryugo and Hohashi 2007; Saotome and Kimura 2011). In the following sections, we will examine whether participants developed "regional attachment" and "self-efficacy" through the FPBL program. For each indicator, items were developed with reference to previous studies (Suzuki and Fujii 2008; Sherer et al. 1982).

First, each item of the Likert-scale question was tabulated and the Mann–Whitney U-test, a nonparametric test, was conducted to determine if there were statistically significant differences between the mean points of indicators in pre- and postconditions. Then, a factor analysis was conducted to extract common factors behind the responses to each item based on a polychoric correlation matrix. Considering the small sample size of the survey, we adopted the least-squares method for the factor analysis, which allows for easy convergence with a small sample size. To determine the appropriate number of factors, we conducted parallel analysis (Thompson 1996), a method used to estimate the error contained in the data, extract the number of factors that have more information (meaningful) than the error, and identify the largest number of meaningful factors. In our analysis, the number of factors assumed was 3 and 2 for the regional attachment and self-efficacy indexes, respectively. In addition, we compared the average of the factor scores between the

Table 21.4 Question items and indicative statements used in the survey

	Pre	Post
1	Please write something you are looking forward to during your visit to Iitate village	Please write about your impressions and something you thought about during your visit to Iitate village
2	Please describe as much as you know about the current charms and challenges of Iitate village	
3	Please answer the following questions. Please select one answer that best describes your opinion, with "7. Agree very much" as the highest rating and "1. Disagree completely" as the lowest rating	
	AT_01	I feel like I have a place for myself in Iitate village
	AT_02	I think it's easy to live near Iitate village
	AT_03	I want to continue to live near Iitate village for a long time
	AT_04	I can relax in Iitate village
	AT_05	I like the atmosphere and local character of Iitate village
	AT_06	I like Iitate village
	AT_07	There are things in Iitate village that would be sad to see go
	AT_08	I think Iitate village is important
	AT_09	There are things in Iitate village that I do not ever want to see change
4	Please answer the following questions. Please select the answer that best describes your opinion, with "7. Agree very much" as the highest rating and "1. Disagree completely" as the lowest rating	
	SE_01	I think I can be of some help and contribute to the reconstruction of Iitate village
	SE_02	I think I can work with the local people in Iitate village
	SE_03	I think I can handle most of the problems that arise in the activities related to the reconstruction of Iitate village
	SE_04	I think there are things I can do for the recovery of Iitate village and its future
	SE_05	I would like to continue to be involved in any way I can in activities related to the reconstruction of Iitate village
	SE_06	I will actively try to learn what I don't know about the reconstruction of Iitate village
	SE_07	I would like to work on something that has not been addressed so far during the reconstruction of Iitate village
	SE_08	I would like to be involved in some way in the reconstruction of Iitate village, even if I fail
	SE_09	I will try to find out what I can do to help with the reconstruction of Iitate village and get started on it immediately

(continued)

Table 21.4 (continued)

Pre	Post
SE_10	I will continue to discuss the reconstruction of Iitate village
SE_11	If I participate in activities that are useful and related to the reconstruction of Iitate village, I think I can help with the reconstruction
SE_12	I think we can work with the local people in Iitate village and contribute to the recovery
SE_13	I believe that if we successively solve each operational problem related to the reconstruction of Iitate village, we will be able to move forward with the reconstruction process
SE_14	I think we can contribute to the reconstruction of Iitate village by being continuously involved in activities related to the reconstruction of Iitate village

Table 21.5 Basic information of respondents

Gender	(Male/female)	67/30
Course	(Undergraduate/Master/Doctor)	78/15/4
First visit	(Yes/no)	68/29
Total		97

pre- and postconditions in accordance with the extracted factor structure, as discussed in detail below. For the Mann–Whitney U-test and factor analysis, the statistical analysis software R (ver. 4.1.0) was used, mainly the "psych" package, which is capable of general statistical analysis. Additionally, a summary and introduction of the free-response statements follows the quantitative analysis for discussion. The results are described in detail below.

21.4.1 Basic Information of Respondents and Summary of 2 Indices

Table 21.5 summarizes the information of participants' gender, course type, and whether this was the first visit to Iitate. The gender and course type of respondents in this survey were generally in line with the proportions of the university student population. About two-thirds of the respondents were visiting Iitate for the first time.

Figure 21.7 shows the results of the Likert-scale responses for "regional attachment" (prefixed with AT_ in the figure) and "self-efficacy" (prefixed with SE_ in the figure) as stacked proportional bar graphs. The results of the Mann–Whitney U-test are also shown in the same figure. The results indicate that the percentage of

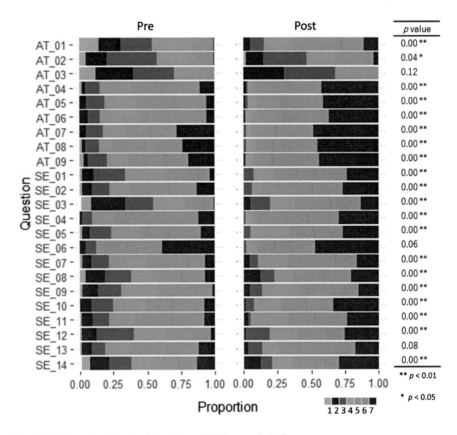

Fig. 21.7 Proportional stacked bar chart of 2 Likert-style indices

respondents who chose options 6 and 7 in the postcondition was generally higher than in the precondition, with no differences found in AT_03, SE_06 and SE_13.

21.4.2   Result of Factor Analysis

Tables 21.6 and 21.7 show the results of the factor analysis for the indicators of "regional attachment" and "self-efficacy," respectively. The number of factors extracted was 3 and 2, and for both results, the explanation rate of the overall data by the extracted factors (Cumulative Var in the table) was about 60%. The structure of the data was easy to understand, with each item having a factor contribution of 0.4 or more to a single factor.

As for "regional attachment," Factor 1, consisting of AT_01, AT_04, AT_05 and AT_06, was named "Preference," because it is a general preference for the target location. Factor 2, consisting of AT_07, AT_08, and AT_09, was named "Cherish,"

Table 21.6 Result of factor analysis on regional attachment index

Item	Factor 1 preference	Factor 2 cherish	Factor 3 livability	Commonality	Uniqueness	Complexity
AT_01	0.54	0.04	0.28	0.55	0.45	1.50
AT_02	0.02	0.04	0.70	0.53	0.47	1.00
AT_03	0.02	−0.04	0.66	0.44	0.56	1.00
AT_04	0.89	−0.07	−0.02	0.69	0.31	1.00
AT_05	0.85	0.06	0.00	0.80	0.20	1.00
AT_06	0.61	0.21	0.04	0.61	0.39	1.20
AT_07	0.13	0.78	−0.07	0.73	0.27	1.10
AT_08	0.06	0.71	−0.04	0.54	0.46	1.00
AT_09	−0.07	0.92	0.07	0.80	0.20	1.00
SS loadings	2.42	2.16	1.11			
Proportion Var	0.27	0.24	0.12			
Cumulative Var	0.27	0.51	0.63			

Table 21.7 Result of factor analysis on self-efficacy index

Item	Factor 1 external	Factor 2 internal	Commonality	Uniqueness	Complexity
SE_01	0.16	0.63	0.57	0.43	1.10
SE_02	−0.10	0.83	0.58	0.42	1.00
SE_03	−0.05	0.76	0.53	0.47	1.00
SE_04	0.11	0.72	0.64	0.36	1.00
SE_05	0.24	0.61	0.64	0.36	1.30
SE_06	0.16	0.58	0.49	0.51	1.20
SE_07	0.55	0.33	0.68	0.32	1.60
SE_08	0.92	−0.18	0.65	0.35	1.10
SE_09	0.71	0.15	0.68	0.32	1.10
SE_10	0.68	0.16	0.65	0.35	1.10
SE_11	0.75	0.06	0.62	0.38	1.00
SE_12	0.78	0.14	0.78	0.22	1.10
SE_13	0.55	0.11	0.40	0.60	1.10
SE_14	0.90	−0.05	0.75	0.25	1.00
SS loadings	5.05	3.60			
Proportion Var	0.36	0.26			
Cumulative Var	0.36	0.62			

because it is a set of consciousness to cherish the subject site. Finally, Factor 3, consisting of AT_02 and AT_03, was named "Livability," because it represents the evaluation of the target area as a place to live.

As for "self-efficacy," Factor 1, consisting of eight items from SE_07 to SE_14, was named "External," because it is a factor related to respondents' efforts and activities toward external environment and people. Also, Factor 2, consisting of 6 items from SE_01 to SE_06, was named "Internal," because it is a set of items related to one's internal attitudes and abilities.

Figure 21.8 shows the results of the comparison of distributions, box-and-whisker plots, and correlation analysis for the scores calculated for the 3 factors of "regional attachment." Considering the probability density plots in the main diagonal of the matrix, a t-test was used for Factor 3, "Livability," which was found to be normal and equally distributed for both pre- and postcondition, and a Mann–Whitney U-test was used to test whether there were differences in scores of the other two factors between the pre- and postcondition. In the first line of the matrix, whisker box plots are shown for each factor scores in both conditions (upper pink box indicates the results of the precondition and lower blue box indicates those of the postcondition) except for the left cell, which indicates the ratio of samples used (100% for both pre- and postcondition). They show that the differences between the factor scores for the pre- and postcondition of all factors were statistically significant ($p < 0.01$). In particular, for Factor 2 "Cherish" (the third line of the matrix), the precondition scores were close to a uniform distribution when referring the histogram in the first

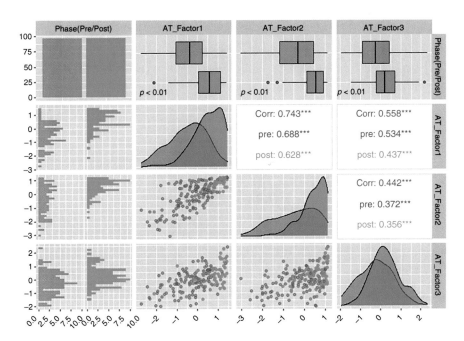

Fig. 21.8 Pairs plot of the factor scores for regional attachment

row and probability density plot in the third row, but the postcondition score shows a keen concentration in the higher level of the score mean. This result indicates that the participants did not show particular tendency for an awareness for this factor, "Cherish," toward Iitate village before participation, but it was sharply enhanced by the end of the program.

Figure 21.9 shows the results of the comparison of distributions, box-and-whisker plots, and correlation analysis of the scores calculated for the two factors related to "self-efficacy." The cell in the middle of the third row (or the scatter plot in the middle of the third line) indicates that there is a strong correlation (for all: 0.743, for pre-condition: 0.716, for post-condition: 0.705 with 1% significant level) between the two factors. Since the probability density plots in the main diagonal of the matrix did not show normality and equal variance between both conditions for both factors, a Mann–Whitney U-test was conducted to determine whether there were differences between the pre- and postcondition scores. The box-and-whisker plots in the first line show different distributions of factor scores for the pre- and postconditions in each factor, and the result of a Mann–Whitney U-test indicates statistically significant difference between conditions ($p < 0.01$). The probability density plot of Factor 1 "External" shows signs of bimodality in the postcondition score, indicating that some participants showed significant changes in this factor whereas others did not.

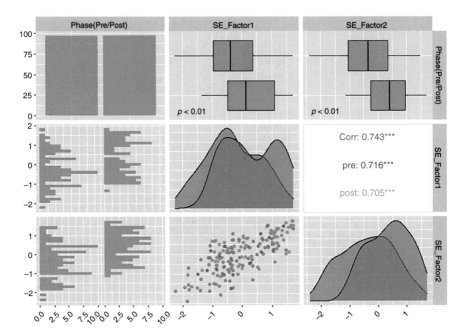

Fig. 21.9 Pairs plot of the factor scores for self-efficacy

21.4.3 Key Qualitative Results from the Free-Style Responses and Writing

Finally, we present here a few key quotes and comments from the participants, excerpted from the result of the free-style responses and reports.

[Male/Undergraduate/First visit]

10 years have passed since the disaster, and I feel that people are starting to lose interest in Tohoku.

[Male/Master's course/First visit]

I felt that tours should be held on a regular basis. It is also necessary to know about and share information on the current situation in Fukushima, including images of the current radiation levels and whether it is safe to live or move there.

[Female/Master's course/second time visit]

I realized that we need to look for a relationship of mutual support that goes beyond the relationship between supporters and recipients.

[Female/Undergraduate/First visit]

People like us who do not live in Fukushima tend to have a fixed impression of Fukushima, and I feel that this may reduce the power of Fukushima's recovery. Therefore, I believe that the easiest way to contribute to the recovery process is to communicate the current state of Fukushima in a concrete way.

[Male/Doctor/First visit]

Through this field study, I felt that it is important to interact with the people of Fukushima. I could also feel the importance of human connections in the reconstruction and resilience of Fukushima. Therefore, I felt that it is important to think about Fukushima together with the local people while interacting with them.

21.5 Discussion

As part of the evaluation of the FPBL program, two psychological indices, "regional attachment" and "self-efficacy," were collected and compared before and after the program as part of a feedback survey of the participants. Participants' scores for place attachment to Iitate after the program experience were significantly higher than those before they participated in the program. In particular, the average score for the "Cherish" factor was significantly higher after participating in the program, suggesting that a positive and sustainable relationship could be formed between the village and the program participants from outside, beyond the learning program anchored to these on-site activities. The overall "self-efficacy" scores also improved after participating in the program. This suggests that the program provided an opportunity for collaboration with the affected areas, which will be necessary on an ongoing basis to pass on the knowledge of resilience and reconstruction to the next generation.

However, it should be noted that AT_03, one of the "Livability" components of the regional attachment factor, did not change after participating in the program. This may be because Iitate still does not have sufficient accommodation and transportation infrastructure for visitors from outside the village and the prefecture. The development of these facilities could be an important foundation for creating and sustaining contact between the village and the outside world when considering the resilience and reconstruction of the disaster-hit area.

In addition, it must be noted that the results for the "External" factor were divided between two groups: those whose scores increased and those whose scores did not increase much. Although the survey alone does not allow us to identify a clear cause for this pattern, the fact that this value tended not to improve in the case of first-time visitors suggests the need for further enhancement of the program's mechanism for welcoming repeat visitors. Further, identifying local activities that contribute to the improvement of this factor will be necessary to further improve the program.

21.6 Conclusion

This chapter presented a multifaceted FPBL framework and specifics that have been developed and practiced in Iitate village, a site that has experienced 11 years of postdisaster reconstruction. The results of the qualitative and quantitative evaluation on the feedback from the program participants indicate that they were able to acquire a concrete awareness and recognition of the disaster-affected area by actually visiting Iitate by directly observing or hearing about the current situation and being involved in field projects. They also seemed to develop a certain level of regional attachment to the village and a sense of self-efficacy toward reconstruction.

This practical and educational program seeks to address the challenges to the practical issues that have emerged in the village through the path of reconstruction with the collaborative efforts of residents, municipalities, supporters such as voluntary NPOs, and universities. The issues emerging in the field are constantly changing, and we must face a different kind of challenge than we did immediately after the earthquake 11 years ago or after the evacuation was lifted 5 years ago. Therefore, going forward, it is expected that this program and related activities will continue to change and evolve with the challenges that emerge in each stage of recovery and further development.

Eleven years have passed since the disaster and people are gradually returning home. Therefore, the agricultural field in Fukushima is considered as a very important region, being full of new attempts and challenges in terms of agricultural education using the FPBL framework from a global perspective. It is hoped that even after the area is restored to predisaster levels, it will long continue to be a site of practice and research as an area where reconstruction knowledge is passed on to future generations, and that the development and sustainability of the area will be ensured through a system of continuous involvement and collaboration of diverse groups of people in Iitate village and the Fukushima region.

Acknowledgements This study is funded by the Fukushima Innovation Coast Initiative Promotion Project named "Utilization of reconstruction knowledge from universities and other institutions." In conducting this research, we received a great support from the Iitate village office and the participants from each university, and we would like to express our great gratitude here. Further, we sincerely appreciate the residents of Iitate village, especially Mr. Muneo Kanno, a farmer in Sasu district, and Mr. Yoichi Tao, the representative of the NPO "Resurrection of Fukushima," for their consistent and enthusiastic support and collaboration in our activities in Iitate village.

References

Akui K, Horita Y, Kubota Y (2019) A study on town planning education through night landscape formation in city center. AIJ J Technol Design 25(61):1367–1372

Allina B (2018) The development of STEAM educational policy to promote student creativity and social empowerment. Arts Educ Policy Rev 119(2):77–87

Bandura A (1977) Social learning theory. Prentice-Hall, Inc., Upper Saddle River

Bender WN (2012) Project-based learning: differentiating instruction for the 21st century. Corwin Press, Thousand Oaks

Brown B, Perkins DD, Brown G (2003) Place attachment in a revitalizing neighborhood: individual and block levels of analysis. J Environ Psychol 23:259–271

Ebitani N, Sugino H, Mizoguchi M (2021) Effect of study tour on emotional attachment for radioactive contaminated area—case study of university student tour to Iitate Village, Fukushima prefecture. J Reconstr Agric Sci 1(1):14–27

Harashima H (2016) Toward new paradigm of engineering—proposal of open spiral model. J Inst Electron Inform Commun Eng 99(4):334–338

Hidalgo MC, Hernandez B (2001) Place attachment: conceptual and empirical questions. J Environ Psychol 21(3):273–281

Ishimori M (2004) Community consciousness and citizen's participation in community building: through a development of a community consciousness scale. Jpn J Community Psychol 7(2): 87–98

Low SM, Altman I (1992) Place attachment. In: Altman I, Low SM (eds) Human behavior and environment (advances in theory and research vol 12). Springer, Heidelberg, pp 1–12

Maeda J (2013) STEM + Art = STEAM. STEAM J 1(1):1–3

Mizoguchi M (2012) New approach for decontamination of farmland in Iitate Village. Trends Sci 17(10):52–56. https://doi.org/10.5363/tits.17.10_52

Mizoguchi M (2013) Remediation of paddy soil contaminated by radiocesium in Iitate Village in Fukushima prefecture. In: Nakanishi TM, Tanoi K (eds) Agricultural implications of Fukushima nuclear accident. Springer, Tokyo, pp 131–142

Mizoguchi M (2016) Challenges of agricultural land remediation and renewal of agriculture in Iitate Village. Water Land Environ Eng 84(6):469–473

Nishiwaki J, Tokumoto I, Sakai M, Kato C, Hirozumi T, Watanabe K, Shiozawa M, Mizoguchi M (2018) Agriculture and radiation safety education through rice harvesting and field trip in Iitate, Fukushima. Water Land Environ Eng 86(1):31–34

Osada Y (2013) Excursion tour in Fukushima held after the 55th symposium. J Jpn Soc Soil Phys 125:55–57. https://js-soilphysics.com/downloads/pdf/125055.pdf

Redish FE (2012) Teaching physics with the physics suite. http://www2.physics.umd.edu/~redish/Book/

Resurrection of Fukushima (2013) Activity report of resurrection of Fukushima. http://www.fukushima-saisei.jp/report201302_en.html

Ryugo C, Hohashi N (2007) Development of a scale for the parent's efficacy of support to encourage their children having blood sampling: finding from the parents who have a child is cared by kindergarten. J Jpn Soc Child Health Nurs 16(1):40–46

Sakai H, Fujimoto T, Ikejiri R (2020) Effects of project based learning with game elements on students' motivation and career vision. Jpn J Educ Technol 43:81–84

Sakano Y (1986) The general self-efficacy scale (GSES): scale development and validation. Jpn J Behav Ther 12(1):73–82

Saotome H, Kimura K (2011) The relationship between goal orientations and generalized self-efficacy among Japanese college ice hockey players. J Jpn Soc Sports Ind 21(2):179–185

Sherer M, Maddux JE, Mercandante B, Prentice-dunn S, Jacobs B, Rogers RW (1982) The self-efficacy scale: construction and validation. Psychol Rep 51(2):663–671

STEM Education Act of 2015 (H.R. 1020, 114-56) (2015). https://www.congress.gov/bill/114th-congress/house-bill/1020

Suzuki H, Fujii S (2008) Study on effects of place attachment on cooperative behavior local area. Infrastruct Plan Rev 25:357–362

Thompson B (1996) Factor analytic evidence for the construct validity of scores: a historical overview and some guidelines. Educ Psychol Meas 56(2):197–208

Wood DF (2003) Problem based learning. BMJ 326(7384):328–330. https://doi.org/10.1136/bmj.326.7384.328

Yakman G (2008) STEAM education: an overview of creating a model of integrative education. Pupil's attitudes towards technology (PATT-19) conference: research on technology, innovation, design and engineering teaching, Salk Lake City, Utah, USA. https://www.iteea.org/File.aspx?id=86752&v=75ab076a

Printed in the United States
by Baker & Taylor Publisher Services